Energy Storage in Electric Power Grids

Series Editor
Bernard Multon

Energy Storage in Electric Power Grids

Benoît Robyns
Bruno François
Gauthier Delille
Christophe Saudemont

iSTE

WILEY

First published 2015 in Great Britain and the United States by ISTE Ltd and John Wiley & Sons, Inc.

ISTE Ltd
27-37 St George's Road
London SW19 4EU
UK

www.iste.co.uk

John Wiley & Sons, Inc.
111 River Street
Hoboken, NJ 07030
USA

www.wiley.com

Library of Congress Control Number: 2015936235

British Library Cataloguing-in-Publication Data
A CIP record for this book is available from the British Library
ISBN 978-1-84821-611-2

Contents

Foreword

At a time when the energy transition has become a necessity at a global level for climatic reasons and to drive the renewal of more responsible economic growth, this book is valuable in many aspects. It tackles the problem of the main technological barrier blocking the large-scale diffusion of new power generation systems based on renewable energy, such as land- and off-shore wind farms, solar energy and even micro-hydraulics. The focus of this book is energy storage, the only solution capable of smoothing the production of these intermittent energy sources contingent upon the natural conditions of wind and sunlight.

Whereas most books on this subject describe in detail the advances in this or that storage technology, this book is among the few – if not the only one – to address the energetic management of systems combining one or more renewable energy-based sources with storage units, with the goal of harmoniously integrating such systems within existing electric grids. The great virtue of Professor Robyns and his coauthors is that through this book, and in a highly progressive manner, they propose a complete methodology for the development of supervisors for the energetic management of systems combining intermittent energy sources and storage units. These supervisors are controlled by fuzzy logic, and the laws that govern them are based on expert knowledge. Due to this approach, multiple objectives can be pursued simultaneously, and the transitions between different operating modes can perfectly be controlled. Moreover, the whole design process is described down to the details of implementation and experimental validation. Several examples of combinations are given, such as a variable-speed wind turbine generator combined with inertial storage, and an electric grid incorporating wind turbine units and adiabatic compressed-air storage. For the latter example, economic optimization is also among the supervisor's objectives.

This book, then, has incontestable practical significance and will undoubtedly serve as a reference in matters of supervisor design for energy generation systems based on renewable sources and energy and power storage units.

Eric MONMASSON
University of Cergy Pontoise
France
March 2015

Introduction

The storage of electrical energy is a long-standing issue that has been only very partially resolved to date, particularly from an economic perspective. Until now, electricity has been produced mainly on a just-in-time basis from flexible resources (hydraulic and thermal based on non-renewable fuels). The development of renewable energies and the need for means of transport with reduced carbon dioxide (CO_2) emissions have generated new interest in storage, which has become a key component of sustainable development. The aim of this book is to contribute to the better understanding of both existing storage technologies and those that are under development, particularly with regard to their management and economic enhancement.

The objectives of this book are to:

– demonstrate the importance of electrical energy storage within the context of sustainable development in intelligent electrical networks or "smart grids";

– show the various services that electrical energy storage can provide;

– introduce the methodological tools used to construct an energy storage management system using a general and educational approach. These tools are based on causal formalisms, artificial intelligence and explicit optimization techniques, and will be presented throughout the book in tandem with concrete case studies;

– illustrate these methodological approaches via numerous concrete and educational examples concerning the integration of renewable energies into electrical networks.

Chapter 1 introduces the issue of storing electrical energy, which cannot be stored directly in alternating current. This observation has shaped the current electrical system, which is based on electricity that is consumed as soon as it is produced. However, the development of intermittent renewable sources and the movement toward more intelligent electrical networks, particularly in terms of energy distribution, are favorable to the storage of this energy. This chapter will introduce various services that storage can provide to a network, thus contributing to its economic enhancement and the issue of its management. A design methodology for this management based on artificial intelligence will be introduced; this is particularly well adapted to the management of complex systems that include uncertainties with regard to production, consumption and the electrical network, targeting multiple objectives and requiring real-time processing, which constitutes a major challenge for future smart grids.

In Chapter 2, we will provide a concise description of the various electrical energy storage technologies used currently, industrially or in the form of demonstrators. The principal characteristics of these storage technologies will be introduced and compared with each other, and the technologies will be illustrated through several examples.

Chapter 3 will examine the general characteristics of the components making up an electrical system. Transport and distribution network management modes will be presented, with an emphasis on the services necessary for their smooth operation, including ancillary services having to do with voltage and frequency control. The potential contribution of storage to these services will also be discussed. Operators of these networks are obviously directly concerned with these services, and so are energy producers and consumers as well as new actors resulting from the liberalization of the electricity market. Examples of the contribution of storage to the treatment of congestion and the dynamic frequency control in the event of sudden instability in an insular network will be presented.

Chapter 4 introduces fuzzy logic, an artificial intelligence method implemented in the remainder of the book. The basic concepts of fuzzy logic will be applied to the management of an inertial energy storage system that is part of a hybrid wind/diesel generator system powering an isolated site.

Chapter 5 will develop a methodology enabling the systematic design of an electrical supervisor for a system incorporating electrical energy storage. This methodology, which includes graphic steps, does not require mathematical models since it is based on an expertise of the system represented by fuzzy rules. Inputs can be random and supervision can target multiple objectives simultaneously. Transitions are progressive between operating modes, as they are determined by fuzzy variables. Finally, this methodology enables the management of storage via convergence toward a state of charge (SOC) and a limitation of its complexity with a view to real-time processing. It is applied to the association of an inertial energy storage system to a variable-speed wind generator, constituting a system capable of supplying ancillary services or functioning in stand-alone mode. We will use an experimental application to discuss the real-time implementation of this type of supervisor, and experimental tests will be used to compare different variants of supervisors.

In Chapter 6, the design methodology of an energetic supervisor is applied to a multisource and multistorage system. The multisource facility studied in this chapter is composed of a wind turbine coupled with a foreseeable and controllable source, and with two storage systems possessing different characteristics. The objectives targeted here are for this facility to be able to be incorporated into a classic network by becoming part of the production planning, despite the random nature of the wind turbine source and the associated production forecast error, and to participate in the stability of the network by contributing to frequency control. The design methodology of the supervisor will be tested on different multisource system topologies in order to illustrate its systematic and modular quality. Performance of various topologies will be compared with the aid of quantitative indicators.

Chapter 7 deals with the management and economic enhancement of adiabatic compressed-air storage incorporated into an electrical network with renewable wind energy production. The objective of this chapter is to analyze the economic enhancement and the uses and interest of medium- and high-power storage devices (ranging from several dozen to several hundred megawatts) for an electrical network. A real-time storage supervision strategy intended to maximize services rendered and thus profitability will be developed using the supervisor construction method introduced in the previous chapters. Three variants of supervisor will be compared: one supervisor limited to traditional economic enhancement based on supply and demand planned on one day for the next day, the proposed real-time supervisor based on fuzzy logic and, finally, a Boolean variant of the latter supervisor. Simulation results show an economic storage gain that is of great significance if it is part of the system services with real-time management.

In the examples examined in this book, the dimensioning characteristics of storage systems (power, energy and dynamics) are assumed to be predefined. The parameters corresponding to these characteristics can be optimized in the same way as the supervisor's parameters by incorporating energy management, with the objective of reducing this dimensioning and thus the associated cost, due to the intelligence incorporated into the supervisor, which constitutes a challenge to the development of storage in economically viable conditions. The examples presented here can be extended to other types of intermittent renewable sources (photovoltaic, small hydropower, marine, etc.) as well as to other storage technologies. Other objectives can also be taken into consideration, such as the aging of storage systems, in order to control their evolution.

1

Issues in Electrical Energy Storage

1.1. Difficulties of storing electrical energy

The electrical energy vector has been highly developed over the past 150 years, as it is extremely practical to use, is not pollutant during use and can generate very little pollution if produced from renewable energies. Its transport over long distances at very high voltage is possible due to transformers, which make it possible to adjust the amplitude of voltage and current waves at will. This possibility offered by transformers goes a long way toward explaining why electrical grids have been developed using alternating voltages and currents.

The weak point of the electricity vector is that electrical currents cannot be stored directly. It is possible to store electrostatic energy (in capacitors) or magnetic energy (in superconducting coils), but the storage capacities of these solutions are quite limited. In order to obtain substantial storage capacities, electrical energy must be transformed into another form of energy. Storage in the form of potential energy by means of turbine pumping stations enables large quantities of energy to be stored, but these stations must be located in regions able to provide significant differences in height between two hydraulic storage tanks. Electrochemical storage using lead batteries has long been used for onboard applications and emergency power supplies, while the storage of kinetic energy by means of flywheels has been used for several decades for fixed applications such as emergency power supplies and some onboard applications, including satellites.

Electrochemical batteries make it possible to store electrical energy in continuous form. Inertial energy storage is used in machines that are required to operate at variable speed, that is, variable frequency. With electrical grids supplying electricity in the form of alternating voltage and currents at fixed frequencies, the implementation of these storage technologies remained complicated until the advent of electronic power, which was introduced in the 1960s, and is currently used to transform the form and characteristics of currents and voltages at will.

The difficulty of storing electrical energy explains why the management of electrical grids has been designed according to the principle of direct consumption of the electrical energy produced, even when the distance between production and consumption is several hundred kilometers. This approach has evolved slightly in France, with the development of nuclear facilities ideally able to produce constant power, favoring the development of hydraulic storage.

The direct consumption of energy has the advantage of a higher overall energetic yield. In fact, the energetic conversion required for storage causes very different losses depending on the storage technologies used. These losses can range from 10% to 50%, or even more. However, this notion of yield can be put into perspective if the stored energy comes from a source for which the non-stored energy would be lost anyway, as is the case with energy that is wind or photovoltaic in origin.

Finally, note, that electrical energy can be stored and subsequently used in another energetic form. This is the case with hot-water tanks in domestic grids, whose final use is thermal energy and the production of hydrogen via electrolysis. Some loads have a storage capacity enabling the control of the power supply from the electrical grid, as with cold storage in supermarket refrigerators, or storage in the batteries of electrical vehicles.

1.2. Why store electrical energy?

The management of electrical grids is based principally on the direct consumption of the electrical energy produced. As consumption is variable, this approach requires the constant adaptation of production to this consumption. Figures 1.1 and 1.2 show typical domestic and commercial consumer profiles, illustrating the variable character of consumption depending on the time of day, season and type of load.

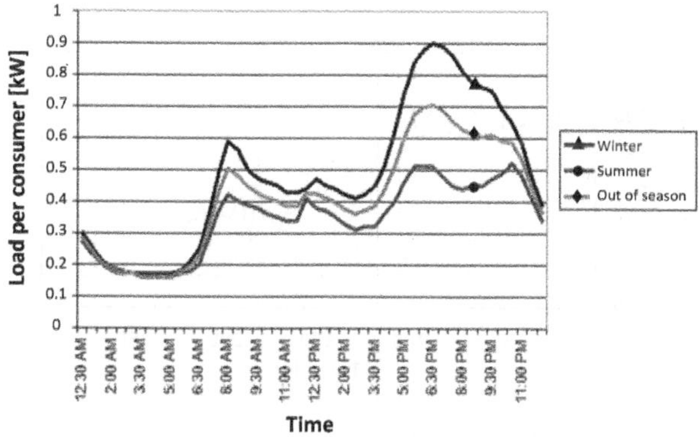

Figure 1.1. *Typical profiles of domestic consumers, not including electrical heating (RTE)*

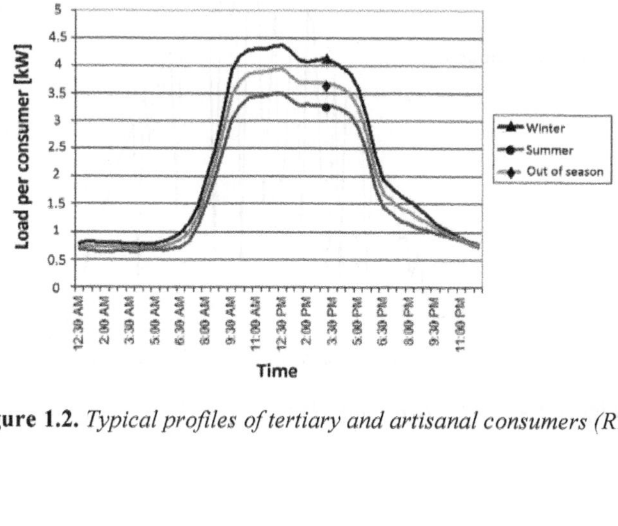

Figure 1.2. *Typical profiles of tertiary and artisanal consumers (RTE)*

Since the development of renewable energy sources, electrical grids have been forced to face the accommodation of highly intermittent production, as is the case for wind, photovoltaic and marine energies, as well as small hydraulic run-of-the-river energies [ROB 12c]. Figure 1.3 illustrates a wind turbine's production of 300 kW more than 5 min. Apart from high variability, fluctuations of 100 kW in 3 s have been recorded. Figure 1.4 illustrates the production of a photovoltaic facility in the span of a day; the presence of clouds induces a high variability of this production.

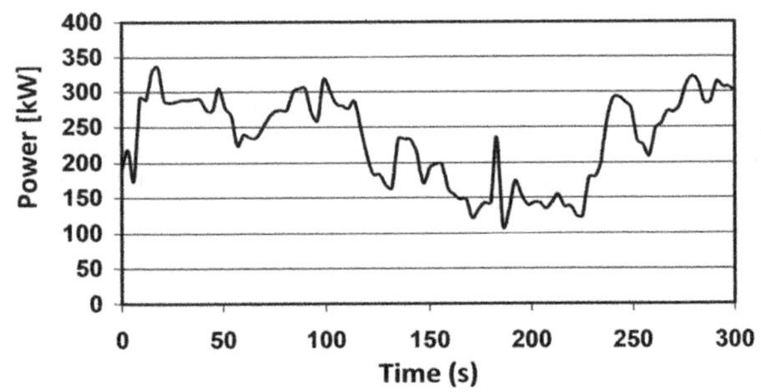

Figure 1.3. *Example of power generated by a fixed speed wind turbine of 300 kW*

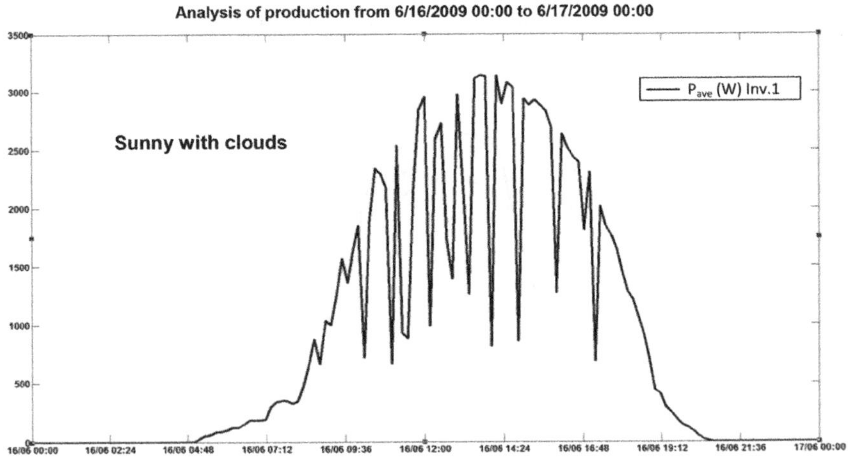

Figure 1.4. *Profile of a sunny day with clouds (source: Auchan)*

Hydraulic resources also show significant fluctuations. For example, ocean waves are an abundant resource, but with large and rapid variations, as shown in Figure 1.5. The flow of a river is also subject to significant fluctuations over months and years, as shown in Figure 1.6, even hours in case of flooding following heavy rainfall. Small run-of-the-river hydraulic facilities, which are not equipped with upstream dams or spillways, will therefore produce uncontrolled variable power when subjected to these fluctuations [ROB 12c].

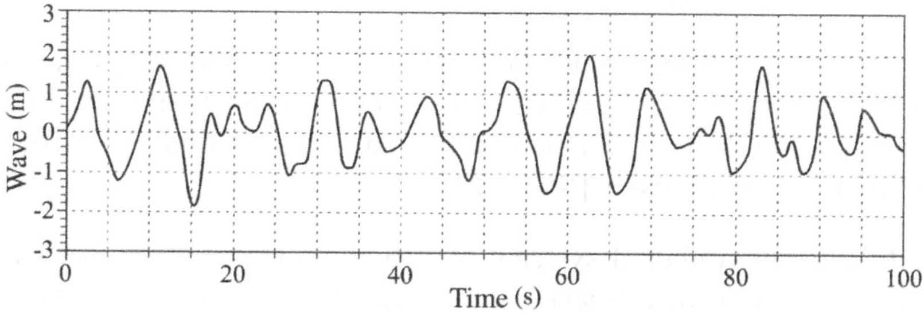

Figure 1.5. *Variation in wave height [MOU 08]*

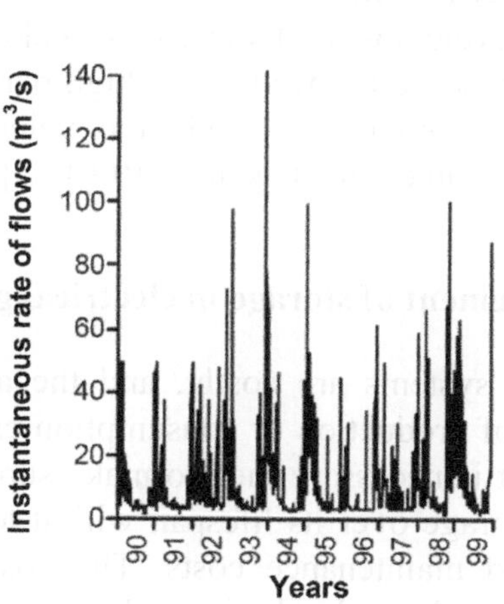

Figure 1.6. *Variations in output of the Oise river over 10 years [ROB 12c]*

These examples show that the balance between production and consumption does not occur naturally, and has been complicated by the increasing development of high-variability renewable energies. Storage of the electrical energy produced by these renewable sources makes it possible to smooth their production, and thus to facilitate their adaptation to consumption.

Conversely, sources such as nuclear power plants ideally produce at constant power. In this case, storage of overproduction during the night makes it possible to compensate for underproduction during the peak hours of the day.

The infrastructures of transport systems such as railways, metros and trams also call grid for fluctuating power on electrical grids due to the starts and stops of traction units, and to fluctuations in traffic at different times of the day [ROB 15].

Finally, the onboard systems of various modes of transport (rail, naval, aeronautical, aerospace, road vehicle, robot, etc.) incorporate electrical storage systems to power backup systems and local electrical grids, recover energy while braking and ensure vehicle propulsion. The development of electric vehicles in particular will significantly increase the need for high-performance onboard electrical storage in order to provide the vehicles with as much autonomy as possible in complete safety [ROB 15].

1.3. Value enhancement of storage in electrical grids

Energy storage systems are costly, and the additional cost they incur in a system of production or consumption can be prohibitive to their installation. It is necessary then to make sure that the economic enhancement of storage over its lifespan will at least compensate for the investment and maintenance costs. The cost of storage varies greatly depending on the technologies and the maturity level of these technologies, which are the subject of a great deal of research and development work. The value enhancement of storage in electrical grids will be dependent on the various services it can provide, which will depend on its positioning in the grid.

Two approaches to the development of storage in electrical grids can be distinguished:

– associated with large intermittent production units (e.g. hydraulic storage associated with wind power connected to the transportation grid);

– diffuse, that is, distributed within the distribution grid, for example.

To make storage profitable, one approach consists of mutualizing the services that a storage system can contribute among various actors (managers, producers and consumers) [DEL 09]. These services consist of:

– local precise and dynamic voltage control;

– support of grid in degraded operation;

– return of voltage in network parts;

– reactive compensation for grid managers (and customers);

– reduction of transport losses;

– power quality;

– energy postponement and support to the production units;

– primary frequency control and frequency stability of insular grids;

– solving of congestion;

– support for participation in ancillary services;

– erasure recovery;

– guarantee of a production profile;

– peak smoothing;

– consumption postponement;

– supply quality/continuity.

The developments presented in this book will illustrate the implementation of several of these services.

The mutualization of services can be associated with a corresponding mutualization of actors; multiple production resources of different types (renewable, difficult to predict and foreseeable, fossil, etc.), multiple consumers and multiple storage systems using different technologies, all with different and complementary characteristics (power, energy and dynamics). Therefore, these are known as multisource, multiload and multistorage systems.

For more than a century, grid management has been based on a centralized approach with limited means of communication, particularly in distribution grids. The implementation and use of new communication technologies along with advanced management resources will increase the intelligence level of grids and contribute to a safe increase in the penetration rate of random productions, while also increasing the energy efficiency of these intelligent grids (Figure 1.7). In this evolution toward smart grids, the storage of electrical energy will play an important role in favoring the development of renewable energies and contributing to the stability of electrical grids, as well as favoring self-consumption in the residential sector, industry and transportation systems. As part of this evolution, the large-scale development of electric vehicles may lead these vehicles to play a particular role, as they represent a significant storage capacity that could contribute to the efficiency and stability of a grid by controlling its load, or even occasionally generating energy on this grid.

The mechanisms of the electricity market also influence the profitability of storage systems. These mechanisms differ from one country to another and, in a competitive environment, evolve with time to favor the development of renewable energies generated on a grid or self-consumed, some loads such as electric vehicles, and energy storage. This storage, connected to an electrical grid, may be seen by this grid as a load or a source depending on whether it stores or generates electricity, which can result in having to pay twice the cost of the grid connection for this device, as a consumer and again as a producer.

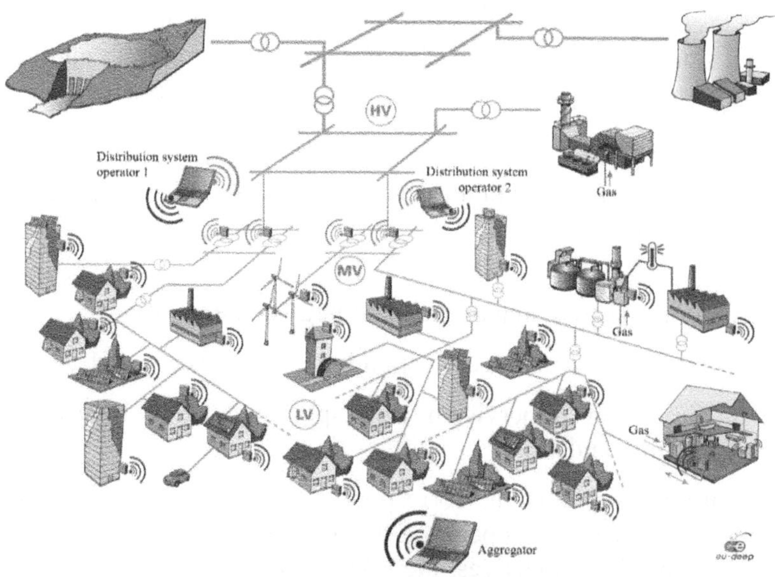

Figure 1.7. *Intelligent grid with communication via internet (source: European eu-deep project)*

The development of energy storage must contribute to sustainable development; therefore, it is important to consider the contribution of these systems to the reduction of CO_2 emissions, not forgetting the gray energy consumed by the construction of the storage system itself. One current trend is the conducting of a lifecycle analysis (LCA) on storage systems.

1.4. Storage management

The management of storage systems in electrical grids must respond to several challenges:

– the development of methods for the supervision of electrical systems whose state or behavior is not well known (random), in which the time horizon to be incorporated may be short (real time when reacting to dynamic stresses) or long (one year, e.g., in order to take into account the seasonal character of renewable sources). These strategies must adapt to an energy policy favoring the expansion of low-power generators dispersed throughout an area, in contrast to the current situation, which is based on the operation of a small number of very high-power production plants (mainly nuclear in France);

– the development of multistorage approaches;

– the development of multiobjective supervision strategies and the pooling of services.

Various time horizons may be put forth in the development of an energy storage system management strategy (Figure 1.8):

– long-term supervision corresponding to a timescale of 1 day;

– medium-term supervision corresponding to a timescale of between 30 min and 1 h;

– real-time supervision, corresponding to the smallest timescale needed to ensure system operation that is sufficient for its stability, the achievement of its objectives, the taking into account of hazards, etc. This timescale can range from a few dozen microseconds to several minutes.

The planning of more long-term storage (several days, weeks, months or years) may also be necessary for the effective management of storage and its economic profitability.

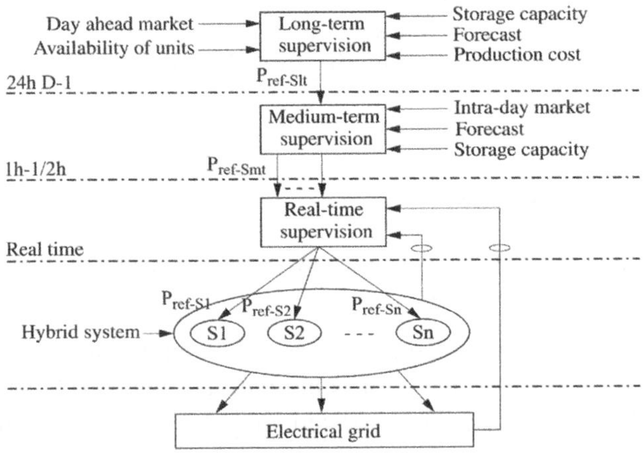

Figure 1.8. *Different time horizons to consider for the management of a hybrid system incorporating one or more sources, storage and possibly controllable loads*

Energy storage management is a significant challenge given the complexity of the problems to be addressed, the economic and ecological objectives and the fact that there is more than one solution that will enable these goals to be met [NEH 11, ROB 12a, ROB 13a, ROB 13b]. Three groups of tools are proposed in the literature to supervise hybrid systems incorporating storage:

– *causal formalization tools* [ALL 10, FAK 11, ZHO 11, DEL 12]. This approach consists of identifying power flows whose inversion can be used to determine reference powers. It requires a detailed mathematical model of the sources and storage systems as well as a good real-time understanding of these different flows and the associated losses;

– *explicit optimization tools* with objective functions [ROB 12b, SAR 13]. This approach is necessary to ensure optimum choice, which guarantees the maximization, for example, of energy produced from a renewable source. The minimization of a well-formulated cost function is difficult to implement, particularly in real time;

– *implicit optimization tools* with, for example, fuzzy logic [CHE 00, LEC 03, LAG 09, COU 10, ZHA 10, MAR 11, MAR 12, ROB 13a, ROB 13b]. This type of tool is well adapted to the management of "complex" systems dependent on the values or states that are difficult to predict and are not well known in real time (wind, sunshine, frequency and states of grid, variation of consumption, etc.).

Different approaches can be considered and combined to ensure storage management: filters, correctors and artificial intelligence technologies.

A design methodology of supervisors dedicated to the management of hybrid energy production systems incorporating storage is developed in this book [ROB 13a, ROB 13b]. This method is an extension of methods widely used in the design of industrial process controls: Petri grids [ZUR 94, LU 10] and Grafcets [GUI 99]. The latter are used to build system controls graphically and step by step in such a way as to facilitate analysis and implementation. They are particularly well adapted to sequential logic systems. However, in the case of hybrid production units that include random variables and continuous states, this type of tool reaches its limits of use. The method proposed is therefore an extension of this graphic approach to include fuzzy values that are not precisely known.

This methodology does not require mathematical models, since it is based on a system assessment represented by fuzzy rules. Inputs can be random and supervision may target multiple objectives simultaneously. Transitions are progressive between operating modes, as they are determined by fuzzy variables. Finally, this methodology enables storage management via convergence toward a state of charge (SOC) and a limitation of complexity with a view to real-time processing.

It can be broken down into eight steps assisting in supervisor design:

– determination of system specifications: the objectives, constrains and means of action must be clearly laid out;

– development of the structure of the supervisor: the necessary supervisor inputs and outputs are determined;

– determination of operational modes by means of functional graphs: a graphic representation of the operating modes is developed based on the knowledge of the system;

– definition of the membership functions of fuzzy variables;

– determination of fuzzy modes by means of operational graphs;

– extraction of fuzzy rules, fuzzy supervisor characteristics and operational graphs;

– definition of indicators used to evaluate the achievement of objectives. These may be, for example, indicators of power, energy, voltage quality, yield or they may be of an economic or environmental nature;

– optimization of supervisor parameters by means of experimental designs and/or genetic algorithms, for example.

This methodology will be progressively developed in this book by considering various applications involving energy storage, based on one technology or a combination of technologies, with renewable wind-power sources and classic sources considered in Chapters 4–7. These examples are readily transferable to the case of photovoltaic sources.

1.5. Bibliography

[ALL 10] ALLÈGRE A.L., BOUSCAYROL A., DELARUE P. *et al.*, "Energy storage system with supercapacitor for an innovative subway", *IEEE Transactions on Industrial Electronics*, vol. 57, no. 12, pp. 4001–4012, December 2010.

[CHE 00] CHEDID R.B., KRAKI S.H., EL-CHAMALI C., "Adaptive fuzzy control for wind – diesel weak power systems", *IEEE Transactions on Energy Conversion*, vol. 15, no. 1, pp. 71–78, 2000.

[COU 10] COURTECUISSE V., SPROOTEN J., ROBYNS B. *et al.*, "Methodology to build fuzzy logic based supervision of hybrid renewable energy systems", *Mathematics and Computers in Simulation*, vol. 81, pp. 208–224, October 2010.

[DEL 09] DELILLE G., FRANÇOIS B., MALARANGE G., "Construction d'une offre de services du stockage pour les réseaux de distribution dans un contexte réglementaire dérégulé", *European Journal of Electrical Engineering*, vol. 12, nos. 5–6, pp. 733–762, 2009.

[DEL 12] DELILLE G., FRANÇOIS B., MALARANGE G., "Dynamic frequency control support by energy storage to reduce the impact of wind and solar generation on isolated power system's inertia", *IEEE Transactions on Sustainable Energy*, vol. 3, no. 4, pp. 931–939, October, 2012.

[FAK 11] FAKHAM H., LU D., FRANÇOIS B., "Power control design of a battery charger in a hybrid active pv generator for load-following applications", *IEEE Transactions on Industrial Electronics*, vol. 58, no. 1, pp. 85–94, January 2011.

[GUI 99] GUILLEMAUD L., GUGUEN H., "Extending grafcet for the specification of control of hybrid systems", *IEEE International Conference on Systems, Man, and Cybernetics*, Tokyo, pp. 171–175, 1999.

[KAN 11] KANCHEV H., LU D., COLAS F. *et al.*, "Energy management and operational planning of a microgrid with a pv-based active generator for smart grid applications", *IEEE Transactions on Industrial Electronics*, vol. 58, no. 10, pp. 4583–4592, October 2011.

[LAG 09] LAGORSE J., SIMOES G.M., MIRAOUI A., "A multiagent fuzzy-logic-based energy management of hybrid systems", *IEEE Transactions on Industry Applications*, vol. 45, no. 6, pp. 2123–2129, November–December 2009.

[LEC 03] LECLERCQ L., ROBYNS B., GRAVE J.M., "Control based on fuzzy logic of a flywheel energy storage system associated with wind and diesel generators", *Mathematics and Computers in Simulation*, vol. 63, pp. 271–280, 2003.

[LU 10] LU D., FAKHAM H., ZHOU T. *et al.*, "Application of Petri nets for the energy management of a photovoltaic based power station including storage units", *Renewable energy*, vol. 35, no. 6, pp. 1117–1124, June 2010.

[MAR 11] MARTINEZ J.S., HISSEL D., PERA M.C. *et al.*, "Practical control structure and energy management of a testbed hybrid electric vehicle", *IEEE Transactions on Vehicular Technology*, vol. 60, no. 9, pp. 4139–4152, November 2011.

[MAR 12] MARTINEZ J.S., JOHN R.I., HISSEL D. *et al.*, "A survey-based type-2 fuzzy logic system for energy management in hybrid electrical vehicles", *Information Sciences*, vol. 190, pp. 192–207, 2012.

[MOU 08] MOUSLIM H., BABARIT A., *SEAREV: système électrique autonome de récupération de l'énergie des vagues*, Techniques de l'Ingénieur, 2008.

[NEH 11] NEHIR M.H., WANG C., STRUNZ K., "A review of hybrid renewable/alternative energy systems for electric power generation: configurations, control, and applications", *IEEE Transactions on Sustainable Energy*, vol. 2, no. 4, pp. 392–403, October 2011.

[ROB 12a] ROBOAM X., *Systemic Design Methodologies for Electrical Energy Systems*, ISTE, London and John Wiley & Sons, New York, 2012.

[ROB 12b] ROBOAM X., *Integrated Design by Optimization of Electrical Energy Systems*, ISTE, London and John Wiley & Sons, New York, 2012.

[ROB 12c] ROBYNS B., DAVIGNY A., BRUNO F. *et al.*, *Electric power generation from renewable sources*, ISTE, London and John Wiley & Sons, New York, 2012.

[ROB 13a] ROBYNS B., DAVIGNY A., SAUDEMONT C., "Methodologies for supervision of hybrid energy sources based on storage systems – a survey", *Mathematics and Computers in Simulation*, vol. 91, pp. 52–71, May 2013.

[ROB 13b] ROBYNS B., DAVIGNY A., SAUDEMONT C., "Energy management of storage systems based power sources and loads", *Electromotion Journal*, vol. 20, nos. 1–4, pp. 25–35, 2013.

[ROB 15] ROBYNS B., SAUDEMONT C., HISSEL D. *et al.*, *Energy Storage in Transportation and Buildings*, ISTE, London and John Wiley & Sons, New York, forthcoming, 2015.

[SAR 13] SARARI S., KEFSI L., MERDASSI A. *et al.*, "Supervision of plug-in electric vehicles connected to the electric distribution grids", *International Journal of Electrical Energy*, vol. 1, no. 4, pp. 256–263, December 2013.

[ZHA 10] ZHANG H., MOLLET F., SAUDEMONT C. *et al.*, "Experimental validation of energy storage system management strategies for a local DC distribution system of more electric aircraft", *IEEE Transactions on Industrial Electronics*, vol. 57, no. 12, pp. 3905–3916, December 2010.

[ZHO 11] ZHOU T., FRANÇOIS B., "Energy management and power control of an hybrid active wind generator for distributed power generation and grid integration", *IEEE Transactions on Industrial Electronics*, vol. 58, no. 1, pp. 95–104, January 2011.

[ZUR 94] ZURAWSKI R., ZHOU M., "Petri net and industrial application: a tutorial", *IEEE Transactions on Industrial Electronics*, vol. 41, no. 6, pp. 567–583, 1994.

2

Recent Developments in Energy Storage

2.1. Introduction

This chapter will provide an overview of the various technologies used to store electrical energy. The maturity levels of these technologies can be very different; some of them have been used for many years and are available on the market, while others are only in the demonstration stage. The principal characteristics of these storage technologies will be introduced and used to compare them with each other. Each of these technologies will be illustrated by several examples.

2.2. Storage technologies

Techniques used for storing energy in fixed applications such as those concerning the production and distribution of electrical energy will be briefly described in this chapter. Storage technologies adapted for onboard systems are obviously more limited for reasons of volume and weight.

The following techniques enable the long-term storage of electrical energy for a period of more than 10 min and up to several months:

– hydraulic pumping and gravity-flow storage, used on a large scale in electrical grids;

– thermal storage in the form of sensitive heat (with no change of state) or latent heat (with a change of state);

– storage in the form of pressure energy by means of compressed air;

– electrochemical accumulator cell batteries available in various types;

– hydrogen storage obtained via electrolysis and the use of a fuel cell to recover the electricity.

The following techniques enable the short-term storage of electrical energy, for 1 s to several minutes:

– storage of kinetic energy in a rotating mass called a flywheel;

– storage of magnetic energy in superconducting coils (superconducting magnetic energy storage – SMES);

– storage of electrical energy in supercapacitors.

Figure 2.1 summarizes the main functions present in a generic electrical energy storage system [MUL 13]. This synoptic shows an intermediary energy form representing the reservoir, that is the part that "really" stores energy, or rather the part that corresponds to a highly reversible change of its internal state. One or more interface converters with the electrical environment are, therefore, necessary to carry out transfers during loading (storage) and discharging phases.

An electronic power converter is often required to properly adapt the form of electrical energy from the aforementioned interface converters to the form necessary for the operation of the overall electrical system; this is usually a source of sinusoidal alternating voltage at fixed frequency (Alternating Current (AC)) or a continuous source at a sensibly constant value (Direct Current (DC)). It plays a major role by providing with an excellent efficiency the conversions necessary, for example, to enable the operation at variable speed of an electric machine, or the charge and discharge at variable voltage or current, respectively, of a condenser or an inductance.

Finally, an electronic control system is necessary for providing numerous vital functions such as conversion control (regulation of machine torque at variable speed, of current, and/or voltage), as well as guaranteeing safety (monitoring, balancing of elements in series, etc.), and above all informing the user about the state of charge (SoC) or, as is increasingly preferred, the state of energy (SoE).

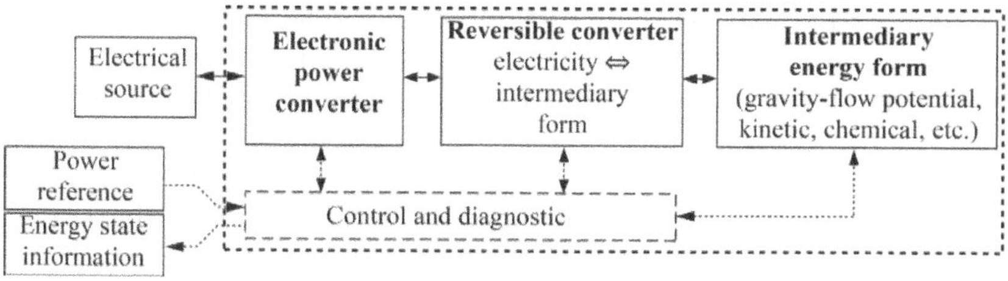

Figure 2.1. *Synoptic of components of a (reversible) electricity storage system as well as its control and diagnostic [MUL 13]*

It is clearly evident that the sizing of a (complete) storage system includes one part associated with energy capacity and another with maximum power (sometimes asymmetric between charge and discharge).

2.3. Characteristics of a storage system

This section presents the main characteristics of a storage system according to the classification proposed by Multon *et al.* [MUL 13]. These characteristics are used to compare the various technologies proposed by builders and to guide the choice of a storage technology for a given application.

2.3.1. *Energy storage capacity*

For an energy storage system, the most fundamental characteristic in theory is its energetic capacity expressed in joules or kilowatt hours. It is a heavily preponderant criterion for the dimensioning of energy storage systems. The amount of energy that is genuinely exploitable depends, however, on the possibility of exploiting all of the energy stored (limits of depth of discharge), as well as on energy losses.

2.3.2. *Maximum power and time constant*

The maximum charge or discharge power (which are sometimes asymmetric) represents another important characteristic, since it qualifies the maximum energy output of a system's performance. It has a very strong impact on dimensioning in power storage systems.

At a given energy capacity, increasing the maximum power requires increasing the dimensioning of certain parts of the system, notably the electronic power converters and those that interface with the physical or chemical value stored (for example, electrode surfaces on electrochemical accumulators). Size, mass and cost are, therefore, affected by this characteristic.

Storage technologies can also be characterized by the ratio of energetic capacity to maximum power; this ratio is sometimes called the *time constant* or the minimum charge/discharge time. With a certain abuse of language, these are called power storage systems when the time constant is small (e.g. less than 1 h) and the power dimensioning impact is high compared to that of energy, and called energy storage systems in the opposite case.

2.3.3. *Energy losses and efficiency*

The various energy transformations that occur during the conversion process are inevitably accompanied by losses, which depend greatly on the technology being considered but can be grouped into two categories:

– First, charge and discharge losses can be highlighted; they are often, at first approximation, proportional to the square of the power flowing.

– Second, we can show no-load losses (in the absence of electrical energy flows in one direction or the other), which are also called self-discharge losses. The latter generally depend on the SoE of the system, and increase along with the SoE. In an electrochemical battery, they may be very low, on the order of a few percent per month, while in a flywheel they can be much higher, up to several percent per hour.

These various losses are used to assess the storage efficiency per cycle [MUL 13].

2.3.4. *Aging*

Storage systems aging like any other object are subjected to stresses, particularly thermal stresses. Their complexity (numerous technologies within the system) multiplies the number of types of aging mechanisms, but some of these are often dominant and are used, once identified, in the development of macroscopic aging laws.

Degradations related to aging are manifested by the deterioration of characteristics such as energetic capacity, which is reduced, or losses, which increase; this usually accelerates the degradation process late in the lifespan and leads, if not attended to, to system failure caused by power surges. Indeed, at the same energy flow levels, a reduction in energy capacity increases relative stresses (more intense cycling), and increased losses cause more significant heating, which itself accelerates degradations.

These degradations occur as a function of time, even in the absence of energy exchanges since temperature is heavily involved, and as a function of load-discharge cycles (rapidity, frequency and depth of discharge).

2.3.5. *Costs*

Cost of investment constitutes the most noteworthy part for buyers, but the systems with the lowest investment costs are generally the systems that degrade the most rapidly, particularly in terms of cycling, and/or in which the efficiency is poorest.

Cost is often specified in €/kWh for accumulators with long time constants (energy storage) and in €/kW for those that are dimensioned in power, with short time constants (power storage).

Operating costs including upkeep, maintenance and energy lost during cycles (possibly weighted by the difference of its value between loading and discharge phases) must also be considered.

Thus, aging and losses over the lifecycle are elements vital to the establishment of a complete economic balance. Moreover, in a system of sustainable development, expenditures of primary materials and gray energy, as well as the additional environmental costs from manufacturing to recycling, must also be considered.

2.3.6. *Energy and specific power*

In the context of onboard applications in particular, mass and volume are important characteristics. From this point of view, electrochemical technologies offer the best performances with mass energies reaching 200 Wh/kg, though they are still often inadequate.

Ragone diagrams, which show power and mass energy, are often used to compare technologies with each other and show the energy/power compromise proper to each of them.

Figure 2.2 shows a simplified example comparing several electrochemical technologies and supercapacitors.

When specific power increases, the mass and volume proportion of the converter section also increases, resulting in a decrease of the overall energy density. Note, however, that Ragone diagrams generally do not take into account all of the components of a complete storage system, but rather are limited to its core, notably excluding electronic power converters. Moreover, at a given set of dimensions, an increase in power-related stresses results in greater losses, thus reducing useful energy capacity. The diagram in Figure 2.2 combines technological dimensioning variants as well as the effects of losses, which leads, for a given group, to an energy density that decreases as the power density increases.

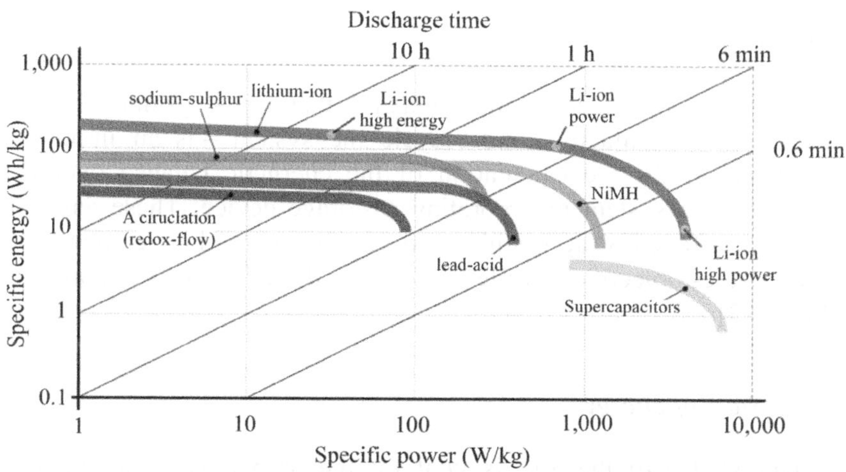

Figure 2.2. *Example of a Ragone diagram for some electrochemical technologies and for supercapacitors [MUL 13]. For a color version of the figure, see www.iste.co.uk/robyns/powergrids.zip*

For stationary applications, the ground area occupied by the whole storage system, sometimes installed in standard containers, is an important criterion. Depending on the application (in energy or power), another significant characteristic is thus the surface energy or power (kWh/m^2 or kW/m^2).

2.3.7. *Response time*

Not all energy conversion phenomena have the same kinetics, and some technologies deliver maximum power more rapidly than others.

In a flywheel system, power can be delivered very quickly since the speed is limited only by the dynamic of the electromagnetic torque of the electrical machine linked to the flywheel, yielding response times in the order of a few milliseconds more or less, depending on scale. Conversely, a turbine pumping station requires an amount of time ranging from 1 min to several minutes to reach full power.

2.3.8. *Gray energy*

Like investment cost, gray energy – that is the primary energy required to manufacture and recycle the system – is an important characteristic for the calculation of the energetic balance over the lifecycle of a system incorporating a storage device. These data are not yet sufficiently accessible, and cannot be used to compare all technologies objectively while considering the effects of aging as well [MUL 13].

2.3.9. *State of energy*

Knowledge of the *SoE* is vital for the proper management of a storage system no matter whatever the application. The SoE is defined as follows:

$$SoE = E/E\text{sto} \quad [2.1]$$

where E is the quantity of energy stored at a given moment; that is the gross energy available if there were no losses at discharge (with the effectively restorable value necessarily being lower depending on the discharge efficiency), and Esto is the energy storage capacity.

An SoE value of 100%, therefore, corresponds to a full charge, and a value of 0% corresponds to a deep discharge (the maximum possible). For various reasons, most storage systems do not tolerate being put in a state of deep discharge due to excessive aging in the case of electrochemical batteries, or simply to a degradation of the maximum power permitted. However, it is also possible that the energy capacity has been determined taking into account the impossibility of achieving a complete discharge.

The evaluation of the SoE is based on an observation of the physical or chemical phenomena implemented.

In a flywheel, the simple measurement of the rotation speed of the flywheel easily yields information on the value of the SoE.

In a supercapacitor or a superconducting inductance, measurements of voltage and current, respectively, give a relatively precise definition of the SoE.

For other "physical" systems, we may cite the water level in the basin of a *Station de Transfert d'Energie par Pompage* (STEP), and pressure in a compressed-air tank.

Finally, electrochemical batteries have all the specific characteristics, but the most reliable method consists of using coulometry, which is the algebraic measurement of the charge combined with corrections in order to take into account the effects of the charge or discharge system and/or measurements of electromotive force when this is adequately sensitive to the SoC. It is the SoC that is usually given, which corresponds to the quantity of charges (in coulombs or ampere-hours) accumulated relative to the nominal capacity, and is similar to the SoE if the electromotive force is relatively independent from the SoC.

At the same time as the indication of the SoE or SoC, we see indicators of the state of health (SoH) or state of aging, particularly for electrochemical batteries; these indicators assess the level of degradation based on estimates and measurements of the drift of parameters such as ohmic resistance or capacity.

2.3.10. *Other characteristics*

Other characteristics can be useful depending on the application, such as those having to do with safety, for example. Any energy storage system has the potential risk for a loss of control, surge reaction, etc. Every technology has its own risks.

Finally, if we consider the massive deployment of storage devices, the rarity of the primary materials used must be considered.

2.4. Hydraulic storage

2.4.1. *Principle of hydraulic storage*

Storage via hydraulic pumps is widely used in electrical grids. 4,200 MW of storage of this type have been installed in France and are known as energy transfer stations by pumping (STEPs). However, this long-term mass storage requires large amounts of space and a significant vertical drop; this is why it was first developed in mountainous regions.

Figure 2.3 illustrates the principle of hydraulic storage. In the storage phase, water is pumped from a lower basin. In the generation phase, this water is turbined by means of a turbine. The same electrical machine is used for both the storage phase (pumping) and the production phase (turbining).

STEPs are the most widespread method of mass storage; their investment cost is one of the lowest available, their per-cycle efficiency is high (65–85% depending on the dimensions) and their lifespans are very long (several decades) [MUL 13]. The first pumping station was constructed in France, on the Lac Noir (Black Lake), in the Vosges Mountains. Numerous mountain STEP sites have been installed in France since then, at Grand'Maison (1,700 MW), Montézic (4 × 220 MW), Revin (4 × 180 MW) and Le Cheylas (2 × 240 MW), and in Belgium at Coo-Trois Ponts (1,060 MW) [BOY 13].

It is also possible to exploit grade separations between subterranean cavities and the ground surface, or between the sea and nearby mountainous basins. Marine STEPs are in operation at Okinawa in Japan, and EDF in France has projects on the islands of Réunion, Guadeloupe and Martinique.

The creation of ring-shaped artificial islands that would be used to store the energy produced by offshore windmill farms is in the development stage. In Belgium, the creation of an island 3 km off the coast near the city of Wenduine is planned; it would extend over a diameter of 2.5 km at 10 m above sea level.

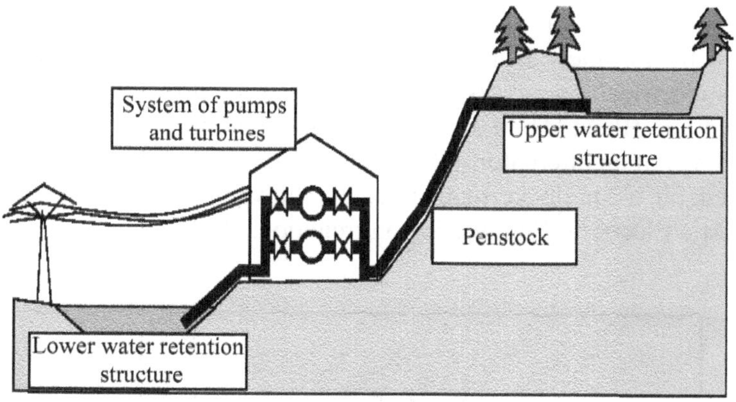

Figure 2.3. *Principle of a pump-turbine station [MUL 03]*

Figure 2.4. *Lac Noir in the Vosges massif and the old pump-turbine station*

2.4.2. *Exercise: Lac Noir station*

Lac Blanc (White Lake) and Lac Noir (Black Lake) are two lakes in the Vosges massif separated by a vertical drop of 120 m. They constitute two reservoirs between which the Lac Noir station causes water to circulate (Figure 2.4). During times of high electricity consumption, water circulates from Lac Blanc toward Lac Noir, driven by turbines. During low electricity consumption times, and when electrical energy is cheap, water is pumped from Lac Noir toward Lac Blanc. The Lac Noir station is equipped with four groups composed of an alternator, a Francis turbine and a pump. The whole is aligned on a single vertical axis, as shown in Figure 2.5 (source: domestic Electrical Engineering aggregation, 1999).

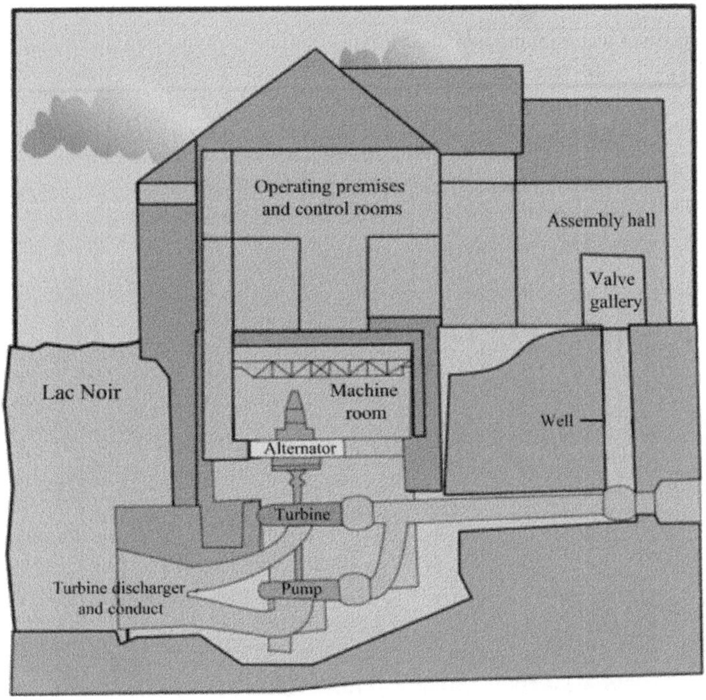

Figure 2.5. *Schema of the old Lac Noir pump-turbine station (Wikipedia, Crochet.david)*

This station, constructed in the 1930s, is currently being renovated, with the station being rebuilt and variable-speed operation enabled.

Principal characteristics of the old station are:

– total available power: 80 MW;

– number of turbine/pump groups: four;

– unitary power: 20 MW;

– maximum height drop: 120 m;

– maximum unitary output of turbines 1–4 Qut: 25 m3/s;

– unitary output of pumps 1–3 Q_{up}: 13 m^3/s;

– unitary output of pump 4: 9 m^3/s.

Lac Blanc has a total capacity of 3,800,000 m^3. The maximum and minimum retainment dimensions are 1,057.6 and 1,041.1 m, respectively.

Lac Noir has a usable volume of 2,000,000 m^3. The maximum and minimum retainment dimensions are 950.5 and 932 m, respectively.

For turbine operation, the designer gives the following characteristics: under a drop of 100 m and a turbined flow of 25 m^3/s, the power supplied by the alternator P_{alt} will be 20 MW.

For pump operation, the characteristics are as follows: for a height difference of 117 m and a flow of 13 m^3/s, the power consumed by the alternator P_{alt} will be 20 MW.

The efficiencies of alternator-turbine and alternator-pump groups are considered independent of drop and discharge heights.

2.4.2.1. *Questions*

1) Calculate, in terms of maximum and minimum dimensions of the two lakes, the maximum height difference H_{max} and the minimum height difference H_{min}.

2) Calculate the efficiency of the group in turbine operation η_t and the efficiency in pump operation η_p.

3) Calculate, for the maximum height difference H_{max} and for the minimum height difference H_{min}, the flows turbined by a group in order to supply 20 MW. Comment.

4) Calculate, for the maximum height difference H_{max} and for the minimal height difference H_{min}, the flows pumped by a group (1–3) which consumes 20 MW.

5) Calculate the electricity consumption in kWh in a pumping system in order to recover 1 kWh using a turbine, with the same height difference.

6) Calculate the theoretical operating time of using a turbine on 2 million m^3 when all four groups consume their maximum unitary flows with a height difference of 100 m.

7) Calculate the theoretical operating time to pump 2 million m^3 when all four groups consume their maximum unitary flows with a height difference of 117 m.

2.4.2.2. *Answers*

1) The maximum vertical drop H_{max} corresponds to the difference in level when Lac Blanc is at its maximum rating and Lac Noir is at its minimum rating.

$$H_{max} = 1057.6 - 932 = 125.6 \text{ m}$$

Likewise, $H_{min} = 1041.4 - 950.5 = 90.9$ m

2) If we use P_h to denote hydraulic power, the efficiencies are determined as follows:

We have $\eta_t = P_{alt} / P_h$ and $\eta_p = P_h / P_{alt}$

In turbine operation, $P_h = Q_{ut} \,.\, \rho_e \,.\, H \,.\, g = 25 \,.\, 1000 \,.\, 100 \,.\, 9.81 =$ 25.525 MW

$$\eta_t = P_{alt} / P_h = 20/24.525 = 0.8155$$

In pumping operation, $P_h = Q_{ut} \cdot \rho_e \cdot H.\ g = 13.\ 1000.\ 117.\ 9.81 =$ 14.921 MW

$$\eta_p = P_h / P_{alt} = 14.921/20 = 0.746$$

Therefore, the overall efficiency of the storage system is $\eta = \eta_t . \eta_p = 0.608$.

Note that modern STEP technologies, particularly those with variable speed, enable higher storage system efficiencies.

3) With efficiency and electrical power being constant, the output is inversely proportional to the height of the drop, or:

At H_{max}, $Q = 25 . 100/125.6 = 19.9\ m^3/s$

At H_{min}, $Q = 25 . 100/90.9 = 27.5\ m^3/s$

With a height difference of less than 100 m, it is impossible to supply 20 MW since the maximum output for a turbine cannot exceed 25 m^3/s.

4) With efficiency and electrical power being constant, the output is inversely proportional to the elevation height, or:

At H_{max}, $Q = 13 . 117/125.6 = 12.11\ m^3/s$

At H_{min}, $Q = 13 . 117/90.9 = 16.73\ m^3/s$

5) Energetic efficiency, which is the ratio of turbined energy to pumped energy, equals $\eta = \eta_t . \eta_p = 0.608$. From this, we can deduce that the electrical consumption for 1 kWh of turbined energy is

$W_{pumped} = 1/(\eta_t . \eta_p) = 1.643$ kWh

6) With the four groups consuming 100 m^3/s, a stored reserve of 2,000,000 m^3 will enable operation for:

$T_t = 2,000,000/100 = 20,000$ s, or 5 h and 33 min

7) With all four groups pumping 48 m^3/s, pumping 2,000,000 m^3 requires a time of:

$T_p = 2,000,000/48 = 41,666$ s, or 11 h and 34 min.

2.5. Compressed-air storage

2.5.1. *Principle of compressed-air storage*

The storage of electricity using compressed air is a procedure that has been known and used on a large scale for more than 30 years. Figure 2.6 shows an illustrative operating schema of a compressed air energy storage (CAES) system. In the storage or load phase, air is compressed in large geological cavities (salt, mining or rocky caves) or in pressurized bottles of gas (on a smaller scale). When the need to produce electrical energy arises (destocking phase or discharge), the air is directed toward a gas turbine driving a generator.

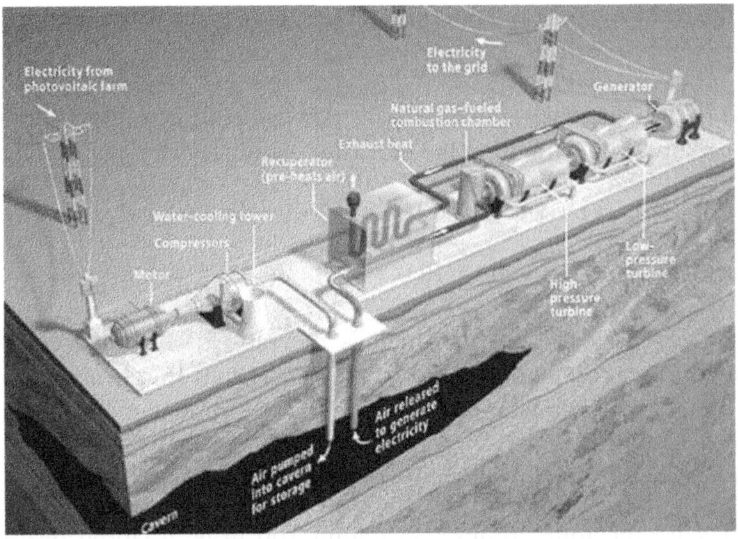

Figure 2.6. *Operating schema of a CAES system [CLI 15]. For a color version of the figure, see www.iste.co.uk/robyns/powergrids.zip*

There are three different types of CAES system technologies: first-generation gas CAES, second-generation gas CAES and adiabatic CAES.

2.5.2. *First- and second-generation compressed-air storage*

The first-generation CAES was set up in Huntorf, Germany, in 1978. Its nominal discharge power was 300 MW, available over 3 h.

The second-generation CAES system was put in operation in McIntosh in the United States in 1991. It supplied a maximum discharge power of 110 MW over 26 h.

Figure 2.7 shows the operating schema of first-generation gas CAES systems.

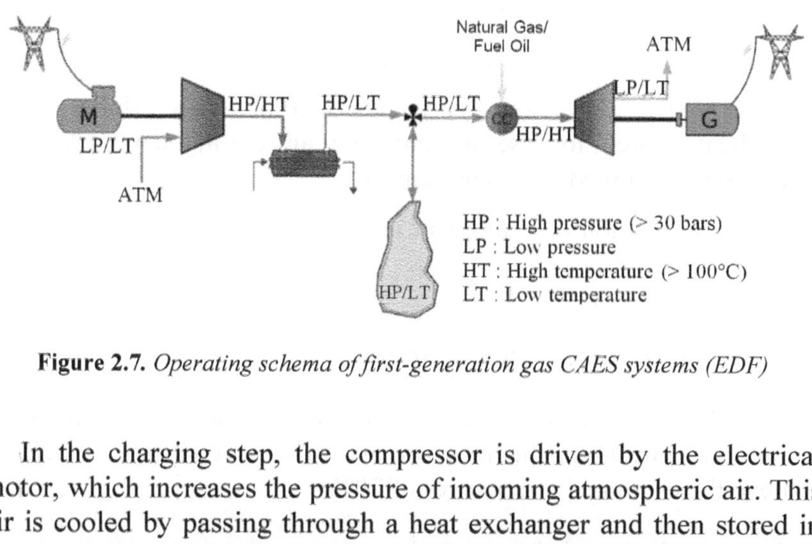

Figure 2.7. *Operating schema of first-generation gas CAES systems (EDF)*

In the charging step, the compressor is driven by the electrical motor, which increases the pressure of incoming atmospheric air. This air is cooled by passing through a heat exchanger and then stored in extremely deep underground caverns. During discharge (the expansion phase), the air is reheated in a combustion chamber using additional natural gas, to be exploited in a machine similar to a gas turbine, which drives the generator. Expected efficiencies from these types of units are between 48 and 50%. To restore 1 kWh to the grid, it is necessary to use around 0.75 kWh of electricity consumed during the pumping phase and to burn 1.22 kWh of natural gas in the combustion chamber.

Conventional CAES can be seen as a way of doing the work of compression with cheap energy, coming from low consumption periods, instead of obtaining it instantaneously through gas combustion. CAES is not a pure storage technology since it is always associated with the use of a fossil fuel (natural gas).

Second-generation gas CAES systems linked to combustion turbines or to any other heat source are concepts still in the processing of reaching maturity. The operating principle of these systems is identical to that of the previous generation, but their designs are different or "hybrid" and are based on modifications of conventional gas turbomachines.

Facilities of this type should be set up in the United States in the years to come and will enable:

– improvement of the overall efficiency to 55%;

– recovery of the energy byproducts of combustion turbines or any other heat source for use in reheating air exiting caverns during expansion (combined cycles principle);

– standardization of equipment used in order to reduce investment costs.

2.5.3. *Adiabatic compressed-air storage*

The latest generation of CAES, called "adiabatic", recovers compression heat via the intermediary of a heat storage system to achieve the estimated efficiencies of 70%.

Its main characteristics are:

– recovery of compression heat using thermal storage, using this energy to reheat air exiting the cavern;

– reduction to zero of polluting emissions during the usage phase, provided that additional gas is no longer used to drive the turbine powering the generator.

Figure 2.8 shows the operating schema for third-generation CAES systems.

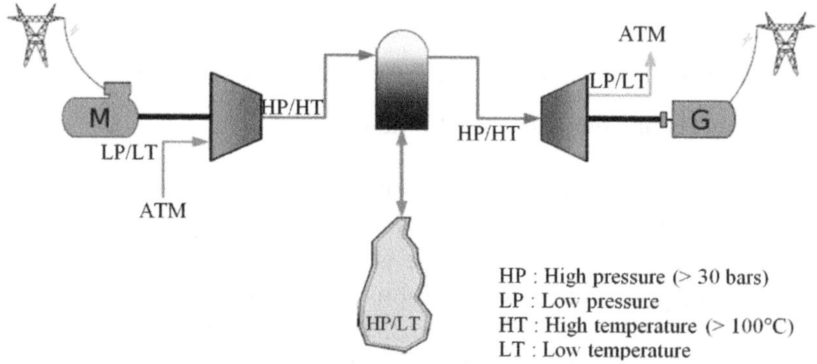

Figure 2.8. *Operating schema of adiabatic CAES systems (EDF)*

In the charging step, the compression process includes thermal storage used to recover heat energy from the air. In the expansion stage, compressed air drives a turbine which powers the generator.

The principal advantages are the reduction of polluting emissions and a significant increase in efficiency. On the other hand, the investment cost remains very high, as this type of technology is still only in the demonstration stage.

2.5.4. *Air storage*

There are three types of solutions possible for the storage of compressed air:

– salt caverns developed specifically for CAES or existing caverns created for salt production and converted for CAES;

– aquiferous or rocky formations;

– mining caverns and caverns dug specifically for CAES, or converted existing caverns: mines, quarries and shut-down underground storage areas.

The most appropriate site is chosen taking geological characteristics into account, including principally:

– a depth of between 200 and 1,000 m;

– thickness of the cavern wall;

– stability of the cavern in the face of pressure changes;

– minerals present and risk of oxidation.

The technique that currently seems the best adapted to CAES, at least from a technological and economic perspective, is the salt cavern dug by dissolution. This technique is used in the production of brine and the storage of liquid hydrocarbons or natural gas. The two CAESs in operation today use salt caverns. Figure 2.9 shows the operating schema of the caverns in Huntorf, which work with a minimum pressure of 46 bars and can store 300,000 m^3 of air, reaching 72 bars.

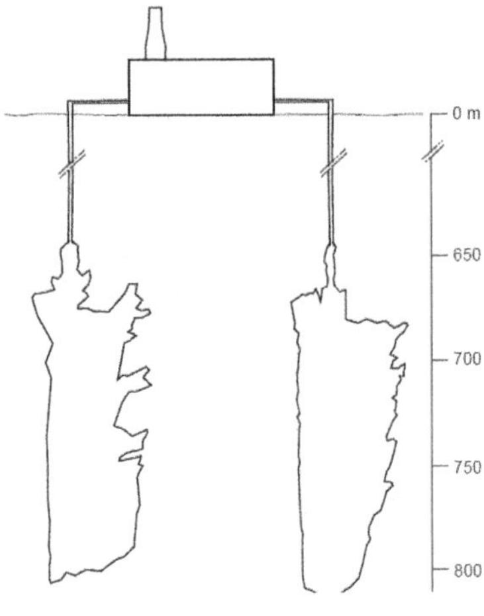

Figure 2.9. *CAES caverns in Huntorf [DAN 12]*

2.5.5. *Hydropneumatic storage*

Hydropneumatic storage (hydropneumatic energy storage (HyPES), or isothermal compressed air energy storage (ICAES)) consists of pressurizing a gas (e.g. air or nitrogen) in a tank by means of a hydraulic motor-pump (Figure 2.10). The use of an intermediary fluid (oil or water) results in a relatively high efficiency; in any case, higher than with an air motor-compressor, as the compression and expansion phases can be quasi-isothermal [MUL 13].

Few systems of this type currently exist. In Switzerland (at the EPFL and the start-up Enairys Powertech), a HyPES2 system with a multi-stage hydraulic motor-pump has been set up with characteristics of 80 kWh–15 kW with a volume of 5.8 m^3. In the United States, the company SustainX set up a 5–1 kWh demonstrator in 2009 and is working on more powerful projects (several MW).

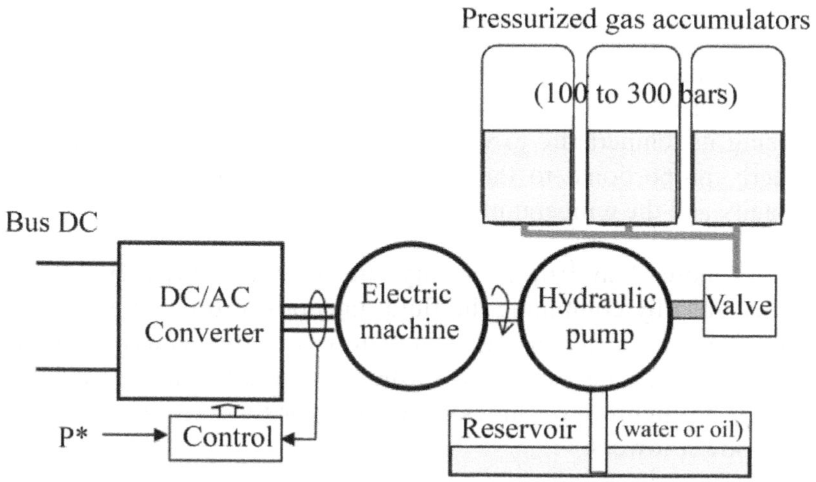

Figure 2.10. *Hydropneumatic storage system with closed air circuit [MUL 13]*

The potentially low cost and very high recyclability of the components make this type of system a serious competitor for electrochemical technologies in stationary applications. Thus, this technology currently seems to be an interesting solution from the cost over lifecycle point of view. However, as yet there are no solutions being sold on the market, though they likely will be in the near future. Improvements in the efficiency–cost–volume are still necessary, as is a better characterization of the cyclical aging of the compressed-air tanks in order to guarantee real profitability over the lifecycle [MUL 13].

2.6. Thermal storage

Two high-temperature thermal storage approaches are possible:

– sensitive-heat storage;

– latent-heat storage.

2.6.1. *Sensitive-heat storage*

This approach is based on the simple physical principle consisting of loading and discharging the quantity of energy in a material and causing its temperature to vary. The quantity of energy stored is thus directly proportional to the mass of the material used, its calorific capacity and the temperature variation.

For more than half a century, the chemical industry has used molten salts by combining the fluid functions of transfer and storage. Sensitive-heat storage using solid materials is also widely used; these are thermal regenerators used in the glass and metallurgy industries employing fireproof ceramics or concrete, or fluidized beds of sand in solar power towers.

In Albuquerque, New Mexico, Areva has introduced an energy storage demonstrator using molten salts in the heliothermodynamic solar park [ROB 12] of the national laboratory Sandia, incorporating concentrators with linear Fresnel mirrors (Figure 2.11). Molten salts are used as a heat-transferring fluid, by extracting a “cold” reservoir (290°C) from them in order to heat them to 550°C on contact with the mirrors, and then passing them through a thermal exchanger to generate the vapor required for the production of electricity. Finally, the molten salts are redirected toward the cold reservoir, and the process can be repeated in a loop; or toward a separate reservoir for storage. The solar power facility can generate electricity both at day and night. Note that in this example there is no reversible storage of electricity, but rather thermal storage assisting with the highly variable production of electricity.

Figure 2.11. *Utilization of molten salts as a heat-transferring fluid in a heliothermodynamic solar power facility (source: Randy Montoya)*

A true electricity storage system can be obtained via thermal pumping (pumped heat electricity storage); this concept consists of using a heat pump to store energy in thermal form in cheap solid materials with a high expected efficiency [MUL 13]. During the load phase, electricity is used to compress the operating fluid and store its heat in a high-temperature container. At discharge, pressurized, high-temperature fluid is used to power a turbogenerator. The efficiency per cycle is estimated at around 70%.

2.6.2. *Latent-heat storage*

Though it presents higher energetic densities, liquid-vapor latent-heat storage is difficult to exploit due to the over-large volume occupied by the vapor phase.

On another level, liquid-solid latent heat (in which the storage capacity is lower, but still much higher than that of sensitive heat) has been studied in great depth for storage in a wide temperature range. In the temperature area studied (on the order of 650°C), only inorganic salts can be used, but the cyclical stability of their properties remains to be assessed.

2.7. Chemical storage

Electricity can be stored electrochemically in non-rechargeable batteries whose use is, therefore, irreversible, as well as in accumulators that can be recharged. Electricity can also be stored in the form of hydrogen obtained via water electrolysis; this process is partially reversible by means of a fuel cell producing electricity from hydrogen and oxygen.

2.7.1. *Electrochemical storage*

Banks of electrochemical accumulators, often called batteries, supply electrical energy to an exterior circuit in the form of a low-voltage continuous current by progressively transforming their internal chemicals following an oxidation/reduction reaction in the electrodes. At the end of the transformation (discharge), the energy storage is emptied. These batteries can be recharged via opposite electrochemical reactions [MUL 13, GLA 12].

Batteries are used mainly in ground transport, particularly in automobiles as starter batteries. Most of these batteries (95%) are lead-acid batteries, but other technologies have been developed, such as cadmium-nickel and lithium-ion batteries, as well as flow batteries, which use different electrolytic couples [BOY 13].

The use of electrochemical batteries in grids has been the subject of large-scale experimentation, and there are several large electricity storage batteries in use.

2.7.1.1. *Lead-acid batteries*

Widely used in the thermal engine-based automobile industry, this kind of accumulator is part of a long-established technology (Figure 2.12). Known for more than 100 years, these batteries remain competitive due to their cost. There are two types: open lead accumulators and gas recombination accumulators. The former have a longer lifespan, ranging from 5 to 15 years. They are cheaper and less sensitive to temperature than the latter, which do not require any upkeep and emit very low quantities of gas. Being the cheapest technology on the market, these accumulators have the disadvantage of supplying a low number of load/discharge cycles (500–1,000 deep cycles) and a relatively low specific capacity (on the order of 30–40 Wh/kg). The largest station, located in Chino, California, has a capacity of 40 MWh and a power of 10 MW. The storage efficiency is typically on the order of 70% [BOY 13], but it is highly dependent on load and discharge systems and can be higher.

2.7.1.2. *Lithium-ion batteries*

Lithium-ion batteries use the circulation of Li+ ions from a negative electrode, generally made of graphite, toward a transition metal oxide (manganese or cobalt dioxide) to generate a current during discharge (Figure 2.13). The advantage of this technology is the mass energy density it supplies (between 80 and 150 Wh/kg), more than five times higher than that of classic lead batteries. In addition, these batteries undergo a relatively low self-discharge compared to other accumulators and require little maintenance. Their expected number of deep load/discharge cycles varies between 1,000 and 4,000, and can be much higher for lower discharge depths. However, their high cost continues to affect their competitiveness negatively. Likewise, recyclability and end-of-life disposal are other areas of study showing notable progress [BOY 13, TES 13, GLA 13].

Figure 2.12. *EnerSys lead batteries: 24 units, nominal voltage of 48 V, nominal current of 200 A, 10 kW power, 1,000 Ah capacity, discharge depth of 72.5% and number of complete cycles estimated at 1,335 (L2EP AMPT Lille)*

Currently, there are three main technologies [COR 13]:

– *Lithium-ion (Li-ion)*, in which lithium remains in the ionic state due to the use of an insertion compound and due to the negative (generally of graphite) as well as positive electrodes (cobalt dioxide, manganese and iron phosphate). Li-ion batteries supply a significant energy density averaging 150 Wh/kg; they have a low rate of self-discharge, no memory effect and require no maintenance. In order to slow down their aging process, it is preferable to use them with a low discharge depth. There is a risk of explosion if they are recharged in poor conditions, which is why designers have developed an adapted security system (battery management system, or BMS) for these batteries.

– *Lithium-ion polymer (Li-Po)* technology, in which the material of the electrodes is identical to lithium-ion technology but the electrolyte is composed of polymer gel. Li-Po batteries are lighter in weight and are safer than Li-ion batteries, but they are also more costly.

– *Lithium metal polymer (LMP)* technology, in which the negative electrode is composed of metallic lithium. LMP batteries have an energy density of around 110 Wh/kg; are entirely solid, which reduces the risk of explosion; and have no memory effect, but optimally function at a high temperature of 85°C.

Figure 2.13. *Saft Li-ion batteries: nominal voltage 48 V, charge current 32 A, discharge current 44 A, energy 3,900 Wh, specific energy 130 Wh/kg, 4,300 cycles at a discharge depth of 80% (L2EP HEI Lille)*

2.7.1.3. *Sodium-sulfur batteries*

Sodium-sulfur (NaS) technology functions by using liquid electrodes. For this, it must be maintained at a temperature of between 290 and 350°C. The electrodes, the site of electrochemical reactions, are made up of liquid sodium and sulfur. The electrolyte separating the two electrodes is made of ceramic, which guarantees good ion conduction. The lifespan can reach 15 years and more than 4,000 cycles in non-critical conditions (discharges lower than 80%).

The sodium-sulfur option can be used for large capacities (several MW and several MWh with a typical time constant of 7 h), which makes it suitable for storage systems supporting electrical grids. NaS batteries have been set up on the island of Réunion (1 MW) and in Texas (4 MW), and numerous applications exist in Japan (for example, 7 units of 2 MW, or 34 MW with a total capacity of 244.8 MWh and an efficiency of 75% associated with a wind farm of 51 MW, Figure 2.14). NaS technology uses widely available, low-cost materials (sodium sulfide, aluminum oxide and aluminum) and is an attractive solution for energy storage in fixed batteries, the regulation of grids and the optimization of station operations over periods of several hours [BOY 13].

Figure 2.14. *Seventeen groups of 2 MW of NaS batteries in Japan [KAW 10]*

2.7.1.4. *Nickel-based batteries*

These are nickel-cadmium (NiCd) and nickel-metal-hydride (NiMH) technologies, both of which supply electromotive force per element of around 1.2 V. Mass performance is up to two times higher than that of lead-acid, and power performance can be excellent [MUL 13].

NiCd technologies, after having initially satisfied the demand for small accumulators for use in electronic applications by the public at large, are now banned from this domain due to the toxicity of cadmium, and are authorized for professional use only. Currently, this technology is typically used in forklifts, and was used in electric vehicles in the final years of the 20th Century. It has also been used as part of a large grid support structure in Alaska [REE 03], where a 1,000-ton nickel-cadmium battery can supply 40 MW for 7 min (4.7 MWh) and 27 MW for 15 min (6.7 MWh).

NiMH technology has partially replaced NiCd technology in the wider public market. It makes it possible to avoid the use of cadmium while increasing mass energy from 60 to 80 Wh/kg and virtually doubling specific energy. It is also used in power batteries in Toyota hybrid drive chains [MUL 13].

2.7.1.5. *Electrolyte circulation batteries*

Circulation batteries (flow batteries) make it possible to circumvent the limitations of classic electrochemical accumulators, in which electrochemical reactions create solid compounds that are stored directly on the electrodes where they are formed. The mass that can be accumulated locally is, therefore, necessarily limited, which in turn limits capacity. As Figure 2.15 [ROB 05a] shows, in electrolyte circulation batteries the chemical compounds responsible for energy storage are liquid, in solution in the electrolyte, and are pumped between the tanks and the electrochemical converter proper (called the stack). The latter is dimensioned in terms of power, while the electrolyte tanks are dimensioned in terms of energy.

Three technologies have been developed to date:

– vanadium and sulfuric acid-based (VRB and Sumimoto);

– zinc bromide-based (several companies, including ZBB);

– sodium bromide and sodium polysulfide-based (Regenesys); the latter technology seems to be facing prohibitive difficulties.

Vanadium technology (Electromotive Force (EMF) of 1.7 V) seems to be among the most promising, with a long lifespan (more than 10,000 cycles) and a very attractive potential cost-to-stored-energy ratio, but one that is linked to the price of vanadium.

High-capacity systems (from approximately 100 kWh to approximately 10 MWh) are in the more or less advanced experimental stage. To give some values, the stack of a VRB system has a power density of 30 W/kg (and 90 W/kg at its peak) for a per-cycle efficiency of around 83% for load/discharge cycles of several hours each, while the electrolyte energy density is around 15 kWh/m^3 [MUL 13].

2.7.2. *Hydrogen storage*

Hydrogen energy storage systems use an electrolyzer. During electricity storage periods, the electrolyzer breaks water down into oxygen and hydrogen according to the equation $2H_2O = 2H_2 + O_2$. This gas is then stored in a liquid, compressed or solid form; in the latter case by the formation of chemical compounds, generally metallic hydrides. There are then three different ways of reinjecting electricity into the grid from the hydrogen [BOY 13]:

– powering a fuel cell, which assumes that the hydrogen will be treated to achieve a certain level of purity;

– synthesizing natural gas via the process of methanization and either injecting it directly into the existing gas network or using it to power a classic gas plant producing electricity;

– finally, using hydrogen directly in a gas plant especially designed for this purpose.

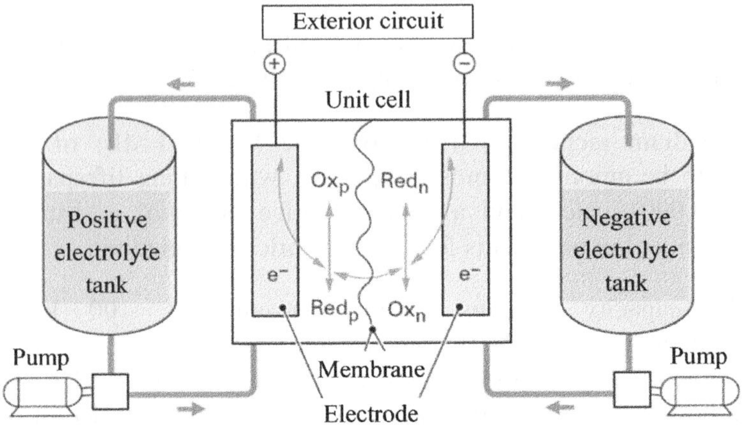

Figure 2.15. *Principle of electrolyte circulation batteries [ROB 05a]*

Hydrogen storage is of interest, as it has a very high energy density and could be used to store large quantities of energy. However, it currently suffers from several disadvantages; a low efficiency from the process, which is around 30% at best; high cost; limited power; short lifespan of electrochemical components; and, finally, the fact that hydrogen poses specific safety issues.

A 200 kW, 1.75 MWh demonstrator was recently set up in Corsica in Vignola, near Ajaccio, as part of the *Mission hydrogène renouvelable pour l'intégration au réseau électrique* ((MYRTE) Renewable Hydrogen Mission for Integration into the Electrical Grid), which is intended to manage optimally the production of a 560 kW photovoltaic station. The objective of the MYRTE program is to develop a system and guidance strategy aimed at improving the management and stabilization of the electrical grid in an island area, in order to exceed the 30% penetration threshold of random renewable energy in non-interconnected areas [BOY 13].

2.8. Kinetic storage

The energy stored in a rotating mass depends on the inertia time J of this mass and its angular speed Ω according to relationship [2.2].

$$E = \frac{J\Omega^2}{2} \tag{2.2}$$

In order to limit its mass, relationship [2.1] shows that, for a given quantity of energy, this flywheel must turn at high speeds. We distinguish between low- and high-speed flywheels according to whether the speed is more or less than 10,000 revolutions per minute (RPM). The limit between these two speed ranges is also expressed in terms of the peripheral speed of the flywheel; the low speed thus corresponds to speeds of less than 100 m/s. When the speed is high, magnetic bearings must be used in order to limit losses by friction depending on speed (Figure 2.16). For this reason, it is also desirable to place the flywheel in a vacuum environment. These constraints make inertial storage systems costly but enable them to have reduced volume and weight, thus making them well suited to onboard applications [HEB 02]. In fixed applications such as those seen in grids, volume and weight constraints are not necessarily critical, which makes it possible to imagine flywheels operating at low speed [BAR 04, ROB 05b] driven by classic electric machines.

This storage system has a very low response time and a long lifespan. It can absorb very great variations in power over very large numbers of cycles. The long lifespan of a flywheel (more than 20 years) and its significant recovery power (1 MW recoverable over 1 h) make it a short-term storage system of interest, well adapted for applications involving regulation, the energetic optimization of a system and the improvement of current quality (reduction of power dips and brief cuts, etc.). Efficiency is high since 80–90% of the energy absorbed can be recovered. The response time is very short, approximately 1 ms, which makes it possible to use this type of storage to regulate frequency on a grid. The technology is reliable and requires little upkeep. Its major disadvantage is the duration of storage, which is limited to 15 min [BOY 13], even though flywheels have been dimensioned for time constants on the order of 1 h (by the company Beacon Power), which shows technical feasibility.

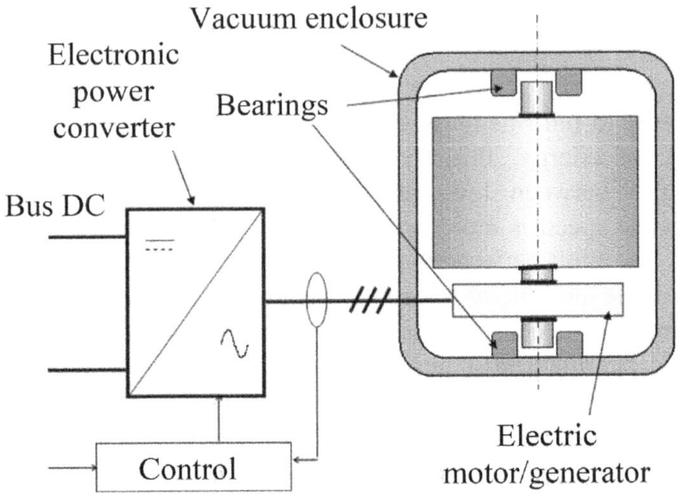

Figure 2.16. *Principle of a kinetic energy storage system [MUL 13]*

2.9. Electrostatic storage

Supercapacitors are the components dedicated to power storage rather than energy storage. They take the form of elemental cells in which storage occurs electrostatically. They have significant mass power, theoretically on the order of 10 kW/kg, supporting between 500,000 and 1 million load/discharge cycles. The principle of supercapacitors is based on the creation of a double electrochemical layer via the accumulation of electrical charges at the interface between an ionic solution (electrolyte) and an electronic conductor (electrode). Unlike batteries, there is no reaction of oxidoreduction. A deposit of activated charcoal on an aluminum film is used to obtain a large surface and thus, a high specific capacity. The electrodes are submerged in an aqueous or organic electrolyte. The electrical charges are stored at the electrode-electrolyte interface. The response time is several seconds [BOY 13]. Several elements with a maximum voltage level of 2.7–2.85 V are placed in series and parallel to each other in order to achieve voltage values and capacities that are exploitable in practice. The supercapacitors available on the market incorporate power electronics ensuring a balance between the charge and discharge of the various elemental units (Figure 2.17), but not the balance necessary to adapt to the continuous bus to which the supercapacitors are connected. The energy stored is dependent on capacity and the square of the voltage at the terminals of the supercapacitor:

$$E = \frac{CV^2}{2} \qquad [2.3]$$

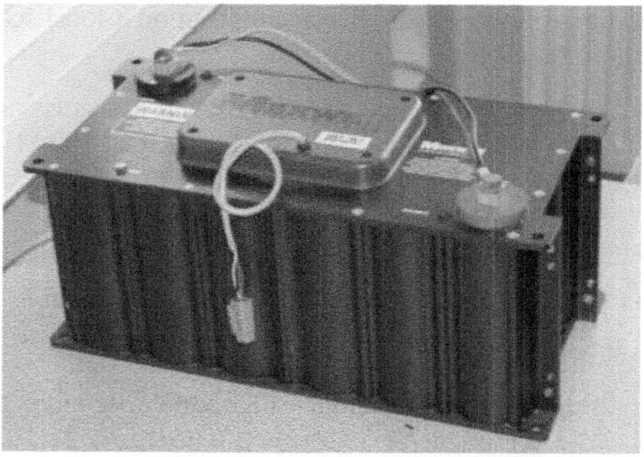

Figure 2.17. *MAXWELL supercapacitor: voltage of 48 V and capacity of 165 Farads (L2EP HEI Lille)*

2.10. Electromagnetic storage

SMES stands for Superconducting Magnetic Energy Storage. Electrical energy is stored via an electric current sent in a superconducting coil cooled below its critical temperature (Figure 2.18). The current circulates almost indefinitely in the coil due to the virtually non-existent losses in the superconducting materials, whose electrical resistance is nil. Thus, SMES constitutes an electromagnetic energy reserve that can be recovered in a very short time (a few seconds or less). The energy stored depends on the inductance of the coil and on the square of the current circulating in the coil:

$$E = \frac{Li^2}{2} \qquad [2.4]$$

According to the quantity of energy stored, we can distinguish between three groups of uses that are possible for SMES on an electrical grid [BOY 13]:

– uninterrupted power supplies (stored energy on the order of several kWh);

– smoothing of production or consumption on a local scale (stored energy of 1–100 MWh);

– stabilization of grids by modulating power transmission (flexible AC transmission systems, stored energy greater than 100 MWh).

For these applications, SMES offers real advantages compared to conventional solutions:

– high efficiency from energy conversion (higher than 85%);

– very short response time;

– long lifespan (a large number of charge/discharge cycles are possible).

The size of SMES is adjustable and can be used to obtain capacities of between 10 kW and 5 MW, with a desired target value of 100 MW. The investment cost constitutes a huge obstacle to the development of this technology, since superconducting materials and cryogenic machinery are still extremely expensive. Moreover, the consumption of energy for the cooling of the superconducting coil reduces the overall efficiency of the installation; those that have been set up to date are mainly demonstration trials [BOY 13].

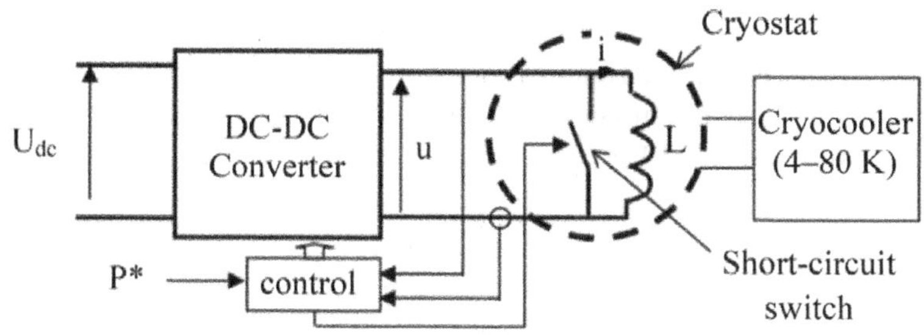

Figure 2.18. *Operating schema of an SMES [MUL 13]*

2.11. Compared performances of storage technologies

Table 2.1 gives the characteristic values of various long-term storage technologies: efficiency, storage density, load/discharge cycle, lifespan and investment costs (CAPEX) in power and energy (CAPEX total = sum of CAPEX power and energy) [MAR 98, BOY 13, MUL 13].

Intermediary energy	Storage system	Efficiency	Storage density kW/m³	Type of cycle	Lifespan	Capex power €/kW	Capex energy €/kWh
Gravity-flow	Hydraulic pumping	0.7–0.85	2 for a drop of 1,000 m	Day, week, season	40–50 years	500–1,500	70–150
Pressure	Air compressor	0.5–0.65	2–5	Day, week	30 years	400–1,200	50–150
Thermal	Latent or sensitive heat storage	0.65–0.85	20–150	Day			
Chemical	Electrochemical batteries	0.65–0.95	5–150	Dozens of minutes, a few days	1,000–12,000 cycles	500–3,000	150–1,200
Chemical	Storage of H_2 via electrolysis and fuel cell	0.25	< 100	Day, season	5–10 years	6,000	< 500

Table 2.1. *Characteristic values of various long-term storage technologies*

Table 2.2 gives the characteristic values of various short-term storage technologies: efficiency, storage density, load/discharge cycle, lifespan and investment costs (CAPEX) in power and energy (CAPEX total = sum of CAPEX power and energy) [MAR 98, BOY 13, MUL 13].

Intermediary energy	Storage system	Efficiency	Storage density kW/m^3	Type of cycle	Lifespan	Capex power €/kW	Capex energy €/kWh
Kinetic	Flywheel	0.7–0.9	10–100	Dozens of minutes	100,000 cycles	500–2,000	2,000–8,000
Electromagnetic	Current in a superconducting coil	> 0.9	0.1–5	Milliseconds to a few seconds	20–30 years	300	> 10,000
Electrostatic	Supercapacitor	0.9–0.95	1–10	Seconds to a minute	500,000 cycles	100–500	10,000–20,000

Table 2.2. *Characteristic values of various short-term storage technologies*

2.12. Bibliography

[ACC 03] "Accumulateur: 40 MW pendant 7 minutes", *Revue de l'électricité et de l'électronique*, vol. 10, p. 8, November 2003.

[BAR 04] BARTON J.P., INFIELD D.G., "Energy storage and its use with intermittent renewable energy", *IEEE Transactions on Energy Conversion*, vol. 19, no. 2, pp. 441–448, June 2004.

[BOY 13] BOYÉ H., "Le stockage de l'énergie électrique, Panorama des technologies", *REE* No.3/2013, pp. 30–41, 2013.

[COR 13] COROLLER P., PASQUIER M., "Les challenges "batteries" et "infrastructures de charge" du véhicule électrique", *REE* No.1/2013, pp. 49–56, 2013.

[CLI 15] climatetechwiki.org/technology/jiqweb-caes.

[DAN 12] DANESHI H., SRIVASTAVA A.K., "Security-constrained unit commitment with wind generation and compressed air energy storage", *Generation, Transmission & Distribution, IET,* vol. 6, no. 2, pp. 167–175, February 2012.

[GLA 12] GLAIZE C., GENIÈS S., *Lead and Nickel Electrochemical Batteries*, ISTE, London and John Wiley & Sons, New York, 2012.

[GLA 13] GLAIZE C., GENIÈS S., *Lithium Batteries and Other Electrochemical Storage Systems*, ISTE, London and John Wiley & Sons, New York, 2013.

[HEB 02] HEBNER R., BENO J., WALLS A., "Flywheel batteries come around again", *IEEE Spectrum*, pp. 46–51, April 2002.

[KAW 10] KAWAKAMI N., IIJIMA Y., SAKANAKA Y. *et al.*, "Development and field experiences of stabilization system using 34MW NAS batteries for a 51MW wind farm", *IEEE International Symposium on Industrial Electronics (ISIE)*, Bari, pp. 2371–2376, 4–7 July 2010.

[MAR 03] MARQUET A., LEVILLAIN C., DAVRIU A. *et al.*, *Stockage d'électricité dans les systèmes électriques*, Techniques de l'Ingénieur, Traité de Génie Electrique, D 4030, May 1998.

[MUL 03] MULTON B., RUER J., *Stocker l'électricité: oui, c'est indispensable et c'est possible. Pourquoi, où, comment?* Publication ECRIN en contribution au débat national sur les énergies, April 2003.

[MUL 13] MULTON B., AUBRY J., HAESSIG P. *et al.*, *Système de stockage de l'énergie électrique*, Techniques de l'Ingénieur, Traité de Génie Electrique, BE8 100, April 2013.

[REE 03] "Accumulateur: 40 MW pendant 7 minutes", *Revue de l'électricité et de l'électronique*, no. 10, p. 8, November 2003.

[ROB 05a] ROBERT J., ALZIEU J., *Accumulateurs. Accumulateurs "redox-flow"*, Techniques de l'Ingénieur, Traité de Génie Electrique, D3357, 2005.

[ROB 05b] ROBYNS B., ANSEL A., DAVIGNY A. *et al.*, "Apport du stockage de l'énergie à l'intégration des éoliennes dans les réseaux électriques. Contribution aux services systèmes", *Revue de l'électricité et de l'électronique*, vol. 5, pp. 75–85, May 2005.

[ROB 12] ROBYNS B., DAVIGNY A., BRUNO F. *et al.*, *Electric power generation from renewable sources*, ISTE, London and John Wiley & Sons, New York, 2012.

[TES 13] TESSARD R., PERRIN M., "Lithium-ion: état de l'art", *REE*, no. 3/2013, pp. 77–83, 2013.

3

Applications and Values of Energy Storage in Power Systems

3.1. Introduction

Whether due to the introduction of new modes of electricity generation using variable renewable energies or to the development of new loads such as electric vehicles, and this within a context of high expectations with regard to quality and price of electricity supplied, power systems are facing greater stresses in terms of planning and real-time management. New uses result in new constraints due to, for example, power-flow reversals that were not originally anticipated in the design of some grids, to an increase in uncertainties to be balanced and to an increase in control dynamics. Due to its capacities to consume or generate electric power on demand, and to the possibility it offers to move energy blocks in time, storage provides a flexibility that seems able to be beneficial for electrical systems and for the development of renewable energies. When connected to a grid, storage can provide services to the various stakeholders in place.

In section 3.2 of this chapter, we will introduce the general characteristics of the components that compose an electrical system. The classic organization of electrical systems developed in the 20th Century is based around so-called "centralized", very high-power generation plants, mainly concentrated near cold sources (thermal power plants) and in regions with favorable natural characteristics (hydraulic power plants). These generation plants are connected to a transmission grid that brings electricity to national consumers and ensures exchanges with bordering countries via interconnections lines. Particularly, for the purpose of reducing currents linked to the routing of this electricity over long distances, thereby reducing Joule losses as well, this grid is operated under very high voltage (HV); in France, the transmission on the national scale is carried out at 400 kV (a level referred to as HV-3 in the French terminology see[1]) or 225 kV (HV-2) and the transportation at the regional scale is carried out at 90 kV or 63 kV (HV-1). The transportation grid is strongly meshed, which enables, among other things, guaranteed continuity of the consumer power supply in the event of an incident such as the unforeseen loss of a line. Large industrial customers such as railway electrical networks are directly connected at the HV level, while most others are powered by distribution grids, which send the electricity over the several dozen kilometers at most which separate each consumer from a primary substation (more than 2,000 in the country). Flowing from a primary substation, electricity first circulates at medium-voltage (MV) level (20 kV or marginally 15 kV in France) and then at low-voltage (LV) level (400 V in France). Unlike transportation grids, distribution grids are not meshed in normal operation. Power switches are used if necessary to reconfigure the MV grid into an emergency operation configuration, which allows limiting the impact of both planned and unforeseen unavailabilities of pieces of grid equipment.

The management of electric power systems developed in the 20th Century is mostly centralized and carried out at the level of the transportation grid to which high-power generation plants are connected. In particular, to follow the variations of the system load

1 In France, the transportation grid operates at HV level (in French "Haute Tension niveau B (HTB)", which designate a rated root mean square (rms) phase–phase voltage $U_n > 50$ kV). Distinctions are made between HV-1 (in French "HTB1": 50 kV $< U_n \leq 130$ kV), HV-2 ("HTB2": 130 kV $< U_n \leq 350$ kV) and HV-3 ("HTB3": 350 kV $< U_n \leq 500$ kV). Distribution grids are operated at MV level (in French "HTA": 1 kV $< U_n \leq 50$ kV) and LV level (low voltage, 50 V $< U_n \leq 1$ kV).

over time, the operating program of these power plants is defined in advance using load forecasting curve, and their reference points are adjusted on an intraday basis if necessary. In order to deal with uncertainties, centralized generation plants also handle services that are vital to system safety and security as well as to the maintenance of voltage and frequency within admissible limits. In this classic operating scheme, distribution grids essentially deal with loads, local generation assets (small hydro, cogeneration, etc.) being highly marginal. They are thus crossed solely by currents flowing from the primary substation to the customers. In these conditions, they are sized so that every consumer can be correctly powered at the peak demand, the on-load tap changer equipping the HV/MV transformers at primary substations is sufficient to control their voltage and the protection scheme is established with only one-directional power flow taken into account.

The development of distributed generation in MV and LV systems considerably modifies the situation. When the installed capacity increases, the power circulating on some grids may become two-directional, which complicates their planning and operation. Moreover, the variable and slightly unpredictable character of certain sources such as windpower and photovoltaics, as well as their behavior when a disturbance occurs, have an impact on the operation of power systems [ROB 04, ROB 06, ROB 12].

The liberalization of the electricity markets within the European Union in the 21st Century has led to the separation of the activities of electricity generation and commercialization, subjected to competition, to the management of transportation and distribution grids. Within metropolitan France, for example, the transmission grid is managed by Réseau de Transport de l'Electricité (RTE), while distribution grids are managed by ERDF in 95% of cases, or by one of the 160 local distribution companies. In this process, the Commission de Régulation de l'Electricité (CRE), an independent administrative authority, has been created to ensure proper operating and compliance with the competitive mechanisms in place. More specifically, for the benefit of end consumers, it handles the surveillance of organized markets and ensures access rights to public grids and the independence of grid managers.

In this new context of growth of intermittent decentralized generation and of liberalization of electricity markets, various technical solutions have been imagined to contribute to the correct operation of power systems with higher shares of renewable energies, such as (1) the construction of interconnection lines, (2) the introduction of new equipment and energy management strategies in distribution grids, (3) the exploitation of control capacities offered by decentralized generators and (4) demand-side management. Among the various possible technical options, energy storage may play a increasingly important role in the future power grids. This trend has already been strengthened in recent years with the availability of new various electrochemical battery technologies integrated in containers, and a progressive reduction in their costs.

This is why, in section 3.3, we will discuss the services that storage can provide to the various stakeholders in power systems from producers to consumers and including grid operators. In each case, in addition to the principles and mode(s) of management that may be appropriate, we will attempt to sketch out the technical requirements and, when possible, define the principles of making a value and even to give some values to assist with simple approaches or lessons from the literature.

This discussion is not meant to be exhaustive, but we hope to provide the reader with an initial understanding of the extent of the subject and its complexity. To go a bit further with regard to certain subjects, we will offer some more in-depth explanations, for example, concerning controls in which the power system security is dependent, and the contributions that storage may make in this area. Later, we will move from these "global" challenges to issues of local constraint. These different applications may enable storage that would impact multiple fields, profit from the "mutualization of services", fully exploit its potential, and thus be more profitable for its operator. Throughout the chapter and at its conclusion, the potential applications of storage will be illustrated by concrete and diverse examples to enrich the panorama offered in these pages.

According to the current grid codes in France, a storage unit can be connected to the transmission grid if its rated power is greater than 12 MW. In this chapter, we will therefore refer to "centralized" storage. For lower power levels, connection can take place in distribution grids and we will refer variously to this as "decentralized", "distributed" or even "diffuse" or "dispersed" storage for facilities of a few kilowatts connected at LV.

3.2. Introduction to power systems and their operation

An electrical system is an ensemble composed of generation units, consumption devices, electric grids and one or more control centers. It may be small in size, as in the case of island grids (Ile de Sein, Guadeloupe, etc.), or may occupy a countrywide scale for larger grids, which can then be interconnected with each other on the scale of whole continents. The development and exploitation of power systems is carried out by various operators, including transmission grid and distribution grid operators. Their main objectives are to guarantee safe and reliable operation, and to fulfill a number of regulatory or contractual commitments while ensuring satisfactory economic performance for local authorities. Figure 3.1 below outlines the organization of an electrical system and locates the various grids, voltage levels and production plants mentioned in the introduction to this chapter.

The sections below give more detail on the components that make up an electrical system, from generators to loads and including transmission and distribution grids. To enable the readers to go a bit further with regard to grid operations, we will then offer a more in-depth examination of the controls that must be organized by transmission system operators in order to ensure the safe operation of electrical systems. Technical details and numerical applications are taken from the French case.

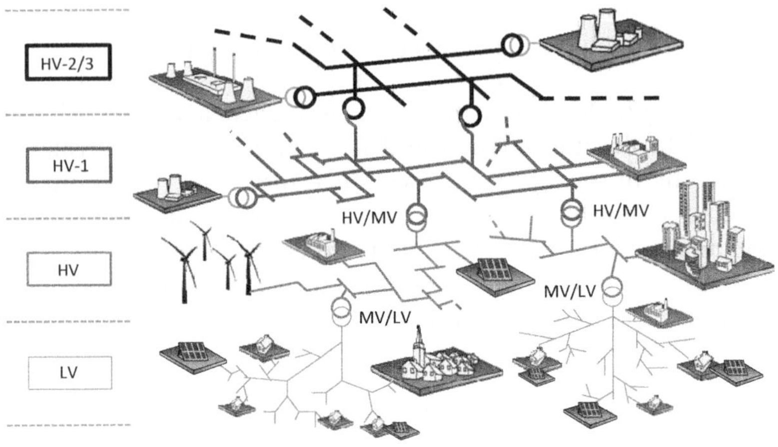

Figure 3.1. *From electricity generation to consumption via transmission and distribution grids: a schematic representation of an electrical system*

3.2.1. *Generation plants*

The production of electrical energy makes it possible to cover the electricity needs of a country or a given region through the transformation of primary energy into electrical energy. Generation plants can be grouped into two categories: so-called centralized generation and so-called decentralized generation. The following sections briefly present the characteristics of these two types of assets.

3.2.1.1. *Centralized generation*

For economic (size effect) or technical reasons (ease of controlling a limited number of high-power plants, specific site characteristics for hydraulic generation, presence of a cold source for technologies involving thermodynamic cycles, etc.), major electric power plants have historically been built in a geographically concentrated manner; this is "centralized" generation. It generally involves plants with high unitary power, connected to the transmission grid [ROB 12].

Electricity producers seek to make their investments profitable. They are working with a view to societal acceptability and in compliance with various commitments, for example, regarding their contribution to ancillary services.

The most widely used technologies are fossil-fired and nuclear power generators, gas turbines and large hydraulic generators. Figure 3.2 illustrates the operating principle of a conventional boiler power plant. In this schema, pressurized liquid water is heated via coal combustion (1) until it is vaporized. It is then sent to the high-pressure turbine (HP, 3), with its outflow controlled by valves (2). The expansion of the vapor in the turbine provides mechanical power. The vapor is then reheated (4) and sent back to the low-pressure and high-pressure (LP and HP) turbines (5). The mechanical power produced by the group of turbines is transmitted to the alternator (6), which converts it into electricity. Finally, by means of the condenser (7) and the cold source (8), the vapor is condensed in order to be reinjected into the boiler.

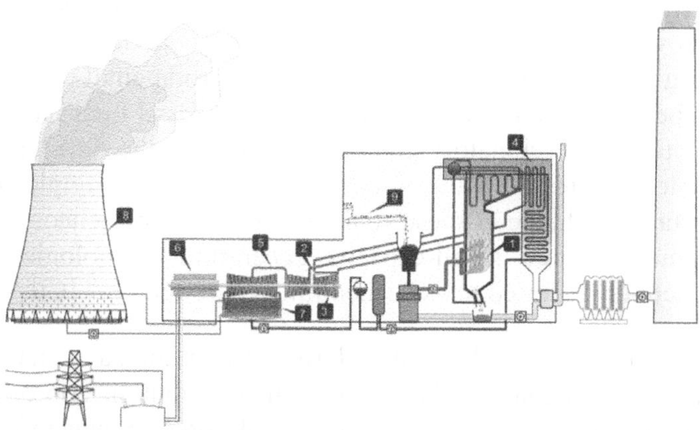

Figure 3.2. *Operation of a conventional boiler power plant: (1) boiler, (2) valves, (3) HP turbine, (4) reheater, (5) LP and HP turbines, (6) alternator (7) condenser, (8) cooling tower and (9) coal [RWE 09]. For a color version of the figure, see www.iste.co.uk/robyns/powergrids.zip*

Nuclear power plants are based on the production of vapor obtained by heating water by means of heat released via a nuclear fission reaction. Like other procedures, they are well suited to a baseload utilization, and this is why pumped hydro storage was developed at the same time in France to optimize their use and manage significant variations in consumption.

Centralized generation plants are interfaced with a control center called a dispatching center, which transmits reference points to them in order to ensure power system security and reliability. In a conventional generation unit, the real power generated is controlled through the admission of the "motive medium" (e.g. via the vapor admission valves in Figure 3.2) and the reactive power is controlled by acting on the field voltage of the alternator. Additional details on these controls will be provided later in this section.

3.2.1.2. *Decentralized generation*

With the liberalization of the electricity market, new investors are developing activities, particularly in electricity production, participation in grid management and demand-side management. However, the construction of centralized generation requires large amounts of money, which is why investors tend to prefer options involving less risk. Moreover, increasing environmental concerns and the prospect of the long-term disappearance of fossil fuels have caused the adoption of a regulatory context favorable to the development of renewable energies [DIR 09]. France, for example, has implemented a peed-in taripp scheme of these sources. All of this has encouraged the development of new small-capacity generation technologies, which are collectively referred to as "decentralized production" [CRA 03].

Ackermann *et al.* [ACK 01] define decentralized generation as electrical sources connected directly to the distribution grid or located downstream of the meter for industrial sites connected to the transmission grid. However, such a definition this remains debatable and plants of several dozen of megawatts connected to the transmission grid are usually considered by the literature as decentralized, as opposed to historic generation, in particular if their primary source is renewable. In France, the voltage range to which a generation facility is connected is defined in the decree of 23 April 2008 and depends on its power: up to 18 kVA in 1-phase LV, 250 kVA in 3-phase LV and 12 MW or even, exceptionally, 17 MW in HV.

The technologies used for decentralized generation vary greatly and can be grouped into two categories: non-renewable energies and renewable energies [CRA 03, ROB 04, ROB 12]. An overview of the main technologies used in practice is presented below.

In France, Windpower and photovoltaics currently represent the majority of new connections in power and in number of sites, respectively. Operators of decentralized generation facilities seek mainly to occupy sites where the resource is most favorable, and to maximize the energy produced, subject to local acceptability and possibilities in terms of connection (costs and time delays).

3.2.1.2.1. Fossil energy sources

Decentralized generation exploiting fossil energy sources such as gas, coal and oil is well established. The principal technologies are combustion or steam turbines and microturbines, combined cycles using both combustion turbines and steam turbines, diesel engines, etc.

More recently, fuel cells have been introduced as an electrochemical device producing electricity and heat via a chemical reaction. Various types of fuel cell exist, but hydrogen is most often used; it can be of renewable origin or not. The advantages of this technology are its efficiency, which can reach 75% in cogeneration, and the fact that it can be used as a means of storing energy [CRA 03]. However, electricity production from fuel cells remains in the demonstration stage for the moment.

3.2.1.2.2. Renewable energy sources

Renewable energies are produced using sun, river or ocean waters, wind, biomass and geothermal energy. The term “renewable” means that these energies renew themselves permanently through time, either continuously or cyclically each day or year or even over several years, as it is the case with tree woods [MUL 03, ROB 12]. In other words, their use at a given time does not threaten the ability of future generations to use them again, unlike fossil sources taken from stocks that reconstitute themselves only extremely slowly (more than millions or hundreds of millions of years) or not at all.

Hydraulic energy is currently the renewable energy that makes the largest contribution to worldwide electricity production [FRI 08, ROB 12], with more than 16% of global production. Biomass is used to generate electricity via the direct combustion of plants or gases yielded by the decomposition of animal or plant products. Geothermal energy uses the heat of the earth. At great depth and high temperatures (more than 150°C), geothermal energy is used to produce electricity, while at lower temperatures (less than 100°C), it is used for heating.

Today, solar energy can be transformed in two various methods in order to produce electricity [SAB 06, ROB 12]. The first, most widely used manner is based on the photovoltaic conversion, which produces electricity directly from the light. The second manner, based on thermodynamics, is used in power plants where the sun's rays are concentrated by mirrors to heat a fluid to a high temperature. The heat from this is then used to produce high-pressure steam, which can then be used to produce electricity in the same way as in a conventional thermal power plant. This method naturally includes storage whose size depends on the quantity of heat-transferring fluid and results from a sizing choice having to do with the desired flexibility of the power plant.

Wind energy uses the force of the wind to produce electricity by means of aerogenerators. The larger the average wind speed of a site, the higher the annual generation [SAB 06, ROB 12].

All generation sites exploiting renewable energies do not have the same characteristics from the point of view of power systems. Some, such as biomass and geothermal energy, are comparable to conventional generation technologies in terms of controllability and of the technologies used to interface them with the grid, in this case synchronous alternators. Others are, among other characteristics, variable and at least partly unpredictable due to the nature of their primary source. This is the case with photovoltaic production, wind production and "run-of-the-river" hydraulic generation. These sources are considered "fatal", which means that the primary energy instantaneously available is lost if it is not used. The increase in this type of intermittent generation is at the root of a greater need for flexibility in the management of power systems at different time horizons. Their association with energy storage systems is often presented by the literature as a means of making their characteristics similar to those of classic controllable sources. This kind of combination may indeed answer certain questions, but will not necessarily dispel other challenges, such as, for example, the weakening of power systems due to the substitution of static converters for synchronous alternators.

3.2.2. *Electric grids*

Electric grids transport energy from generation plants to consumption sites. They are composed mainly of lines, cables and transformation stations combined with protective mechanisms [SAB 07, SAB 08].

Overhead lines are composed of pylons, conductors, insulators and possibly lightning guard cables. The sizing of these components can be grouped into three categories: geometric sizing, mechanical sizing and electrical sizing. Geometric sizing is used to guarantee insulating distances between conductors or in relation to objects located on the ground. Mechanical sizing ensures that the lines will not break due to stresses caused, for example, by the accumulation of snow. Finally, electrical sizing ensures that the temperature reached by the line will not exceed the limit values. The thermal limit is called permanent maximum admissible intensity (IMAP) in France [GAU 97]. Powers involved in heating the line are the power due to the Joule effect and the power due to sunshine, while powers involved in heat dissipation are the power dissipated by convection and the power dissipated by radiation. Thus, the IMAP of a grid may differ according to the season [VER 10]. Insulated cables are conductors surrounded by a sheath that insulates the cable from its external environment. Thus, cable sizing is linked to the power transported, dielectric losses, mechanical properties and capacity per unit length of a cable.

Conductors have an impedance that causes voltage drops depending on the lengths deployed and the amounts of power routed. Maintaining voltage in a grid is the responsibility of the grid operator who, to do this, uses various types of connected equipment (transformers equipped with on-load tap changes, capacitor or inductance banks, generation plants, etc.).

A meshed, interconnected transmission grid ensures the transport of electrical energy between large plants and large consumption centers. It is useful to distinguish between:

– the main transmission grid, intended to transport large quantities of energy over long distances. It supports the maintenance of the overall balance between generation and demand by relying on the interconnection of generation assets with each other. This international mutualization, which was initially done for technical reasons, also serves today as an instrument for commercial exchanges. In France, and in Europe generally, operation under 400 kV was adopted after World War II as the best technical and economic compromise, plus some connections at 225 kV;

– the regional transmission grid, enabling transport over more limited distances, operates at a regional scale. Besides the 225 kV, which directly powers high population density areas, the voltage levels used in France are 90 and 63 kV. Large industrial customers are usually connected directly to these regional transmission grids, as well as medium-size power plants (250 MW or less).

Given the extent of the land areas, French island grids do not include main transmission, but rather certain levels of regional transmission, with the maximal voltage used being 90 kV in Corsica and Guyana, for example, and 63 kV in la Réunion, Guadeloupe and Martinique. Smaller grids, such as those on Saint-Pierre and Miquelon, have no HV voltage level.

Substations are the interfaces between different voltage levels in a grid, and are used particularly to modify the grid's topology in order to enable better distribution of power flows [DEV 09]. As an example, we would cite the primary substations, which count more than 2,000 in France, which can be seen in the distribution grid structural illustration given in Figure 3.3. Besides their main roles as voltage transformers (HV/HV), they provide protection to the grid (amperometric protection, neutral grounding, etc.), measurement of energy flows via their metering equipment, rate changes via centralized remote control at 175 Hz and aid in ensuring electrical system security, particularly via automatic under-frequency load shedding. Source substations contribute to the quality and continuity of the power supply via automatic re-engagement, voltage control and reactive compensation systems.

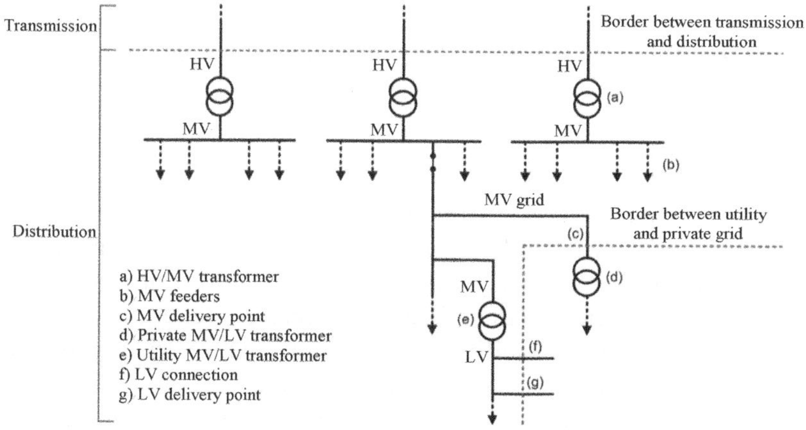

Figure 3.3. *Structure of a distribution grid [DEL 10]*

The HV grid, mainly operated at 20 kV in France, is composed of all of the feeders coming from the primary substations, and supplies power to customers connected at HV and the HV/LV secondary substations. Urban areas are usually connected by underground power lines, while areas with low charge density are powered by overhead or mixed lines (part overhead, part underground). The HV grid is radial and can generally be reconfigured if required to ensure continuity of the power supply, but is non-looped when in operation. The LV grid is composed of the feeders coming from HV/LV substations and supplies power to consumers at 230/400 V.

Transmission and distribution system operators act under regulations to ensure the safety, security and quality of the grids they maintain and develop. In application of regulations and the concession contract (the conceding authorities being the national government and municipalities, respectively), under the control of the CRE, they must guarantee to all users non-discriminatory treatment, the availability of information (transparency) and the confidentiality of sensitive data. Financed via the *Tarif d'Utilisation des Réseaux Publics d'Électricité* (TURPE) (public electricity grid usage rate), grid managers must minimize the economic and environmental impacts of the carrying out of their duties. They are vital stakeholders in the electric system and are interconnected with the majority of its other stakeholders.

3.2.3. *Demand*

Electric consumption corresponds to the electrical energy needs of a region. Consumers seek competitive prices above all, as well as, depending on their type, certain guarantees (stability of prices, green energy source, etc.) and/or various services (energy efficiency, reliability, etc.). Their facilities must comply with various regulatory requirements, which are most often included in grid codes and contracts pertaining to grid connection, operation and grid access. As well as taxes, the price of electricity for consumers includes two components: the first covers the supply of energy and the second has to do with its transmission and distribution, that is, with grids.

Load forecasting is necessary for the operation of a power system and, generally, a region's demand curve is relatively well defined with peak and off-peak hours within a day as shown in Figure 3.4. To follow the temporal fluctuations of the load, generation must be adjusted in order to ensure the balance between demand and supply at all times. Peak times of electricity demand can be distinguished according to the period observed: the daily peak, the seasonal peak in France due to extensive use of electric heating, etc. It is also possible to observe local peaks corresponding to specific needs related to human activity (in industrial areas, tertiary buildings, commercial centers, island areas, etc.).

The large-scale development of wind and photovoltaic generators can make the balancing process more difficult than in the past, as these sources are variable and at least partly unpredictable (being dependent on meteorological hazards); they do not necessarily correlate with the load demand and cannot generally be controlled in real time at present. New situations appear and have an impact on electricity prices, for example, a deficit in generation to manage a peak of 102 GW of consumption on 8 February 2012 quickly drove the market price per megawatt-hour to €2,000. Conversely, the overproduction of electricity during off-peak times can lead to negative prices, which means that producers are forced to pay in order to drain away their electricity.

Thus, storage may, in competition with other solutions, contribute to this balance between generation and demand.

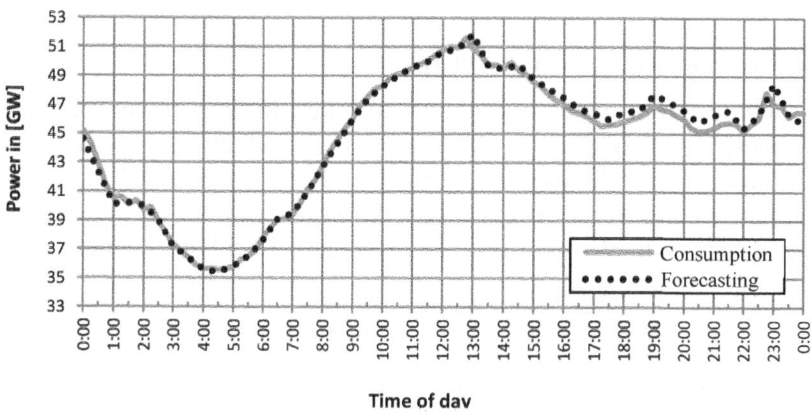

Figure 3.4. *Evolution of the system load in France on 18 October 2010 and comparison with the forecasting made the previous day (D-1)*

3.2.4. *Some basics of the operation of power systems*

The goal operation of a power system is ensured by the transmission system operator, in this case RTE in France. It is intended to accomplish various objectives, including (1) guaranteeing system security, (2) favoring economic performance and the proper proceeding of the mechanisms of the electricity market and (3) fulfilling contractual commitments made to customers, particularly in terms of quality of power supply [RTE 04].

In order to fulfill these objectives in the face of the permanent uncertainties affecting demand (temporal variations, sensitivity to temperature and cloud cover, etc.) and generation (outages, power variability of renewable sources, etc.), various automatic and manual controls with relevant safety margins are applied over different time horizons. These controls are aimed at managing four types of risks liable to lead to generalized power cuts, specifically: (1) the risk of frequency collapse, (2) the risk of voltage collapse, (3) the risk of loss of synchronism and (4) the risk of cascading overload. The subsections below will discuss the mechanisms in place to prevent these four types of risks [ENT 14, RTE 04]. Note that we are examining the French case exclusively; the issues are virtually identical from one electrical system to another, but the technical solutions historically implemented to handle them can differ greatly.

3.2.4.1. *Guarding against frequency collapse*

The design of generation plants (turbines, backup power, etc.) and of some loads takes the value of the frequency into account. Consequently, frequency must be maintained within regulatory/contractual ranges of operation determined on the basis of references such as the European standard EN50160, which gives [EN 99]:

– 50 Hz–1 Hz/+1 Hz for 99.5% of the time over a year;

– 50 Hz–6 Hz/+4 Hz for 100% of the time.

Frequency control is based on the balance between generation and demand and is fixed by the rotaion speed of synchronous alternators, the main technology used for the electricity generation. Remember that the rotation speed of a synchronous alternator is permanently proportional to the frequency of the voltage at its terminals, and the proportionality ratio depends on a design parameter called the number of pole pairs. As the frequency is identical at every point of a grid in a steady state, the various synchronous generators connected to it are joined with each other. Equation [3.1] gives a schematic representation of the equation of the rotating masses describing the variation in alternator speed (Ω_t) according to the variations in the total mechanical torque of generation plants (T_p) and the torque exercised by consumption and losses (T_c):

$$J_s \frac{d\Omega_t}{dt} = T_p(t) - T_c(t) \qquad [3.1]$$

where J_s is the total inertia of the electrical system, which is the sum of the inertias of all connected generation plants. Due to its large value, it constitutes an energy storage system in an inertial mechanical form. This equation can be rewritten by bringing in the total power produced (P_p) and the total power consumed (P_c), with any difference being instantaneously added or subtracted from the total energy of the rotating masses of the system ($E_t = \frac{1}{2} J_S (\Omega_t)^2$):

$$J_s \frac{1}{2} \frac{d(\Omega_t)^2}{dt} = \frac{dE_t}{dt} = P_p(t) - P_c(t) \qquad [3.2]$$

Starting from a steady state, any reduction in generation or increase in consumption makes the right-hand term of the two preceding equations negative: in these conditions, the rotating masses of the electrical system release kinetic energy. This induces a reduction in the rotation speed of the alternators, and thus in system frequency. Conversely, starting from steady state, any increase in generation or reduction in consumption is inducing an increase in frequency. For a given imbalance, the rate of change of frequency is inversely proportional to the system inertia.

In order to limit frequency variations within an operation area where system security is guaranteed, it is necessary to maintain the generation-demand balance by continuously adjusting the operating points of the generation plants, or the power consumed by controllable loads. In practice, in France, three levels of power/frequency control with a given volume of control reserve and a given release dynamic are carried out (Figure 3.5, Table 3.1). First, primary frequency control automatically re-establishes the balance between generation and demand in the seconds following a disturbance such as the outage of a large generator. Next, secondary frequency control automatically takes the frequency back to its reference value within a few minutes and cancels out the differences in exchange on interconnection lines in relation to the planned values. Finally, tertiary frequency control is basically the set of manual actions undertaken in a centralized manner by the system operator within the dozens of minutes or hours after an incident (modification of the operating points of generators, shutdown or start-up of generation assets, etc.).

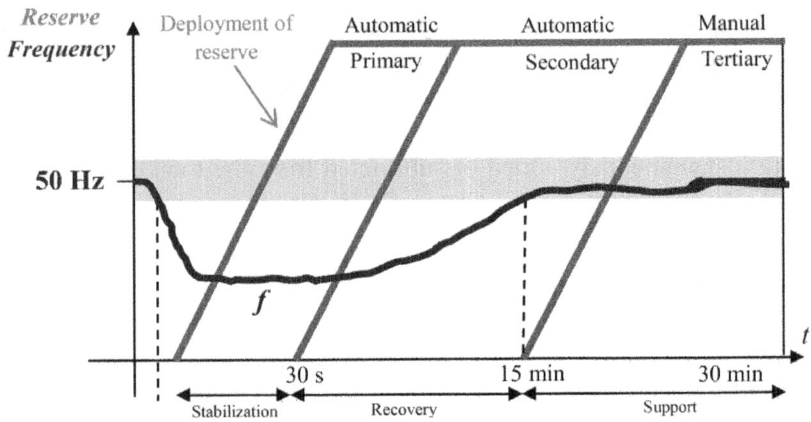

Figure 3.5. *Deployment of primary, secondary and tertiary reserves*

Control	Variation in active power	Deployment time	Duration of delivery
Primary	Proportional to the frequency deviation	<30 s	At least 15 min
Secondary	According to a level of remote control	From 2 to 13 min between $N = -1$ and $N = 1$	As long as necessary
Tertiary	Upon request by the system dispatching	1,000 MW in 13 min maximum	As long as necessary

Table 3.1. *Some technical data on frequency control*

3.2.4.1.1. Primary frequency control

During a major power imbalance, it is necessary to quickly re-establish the balance between generation and consumption in order to avoid the risk posed by a significant variation in frequency. This is the role of primary control, which must be activated quickly as the frequency drops in the first seconds of the transient are critical, as shown in Figure 3.5. In practice, this control is implemented locally on the speed controller of the generating unit because of the required short response time. As can be seen in Figure 3.6, the speed controller acts on the motive medium inflow of the generation unit (vapor, fuel, water flow, etc.) according to the rotation speed of the alternator, i.e. according to (steady state) frequency. The control achieved is proportional; starting from the predetermined reference point, the power generated is increased in the event of underfrequency and reduced in the event of overfrequency (see Figure 3.7).

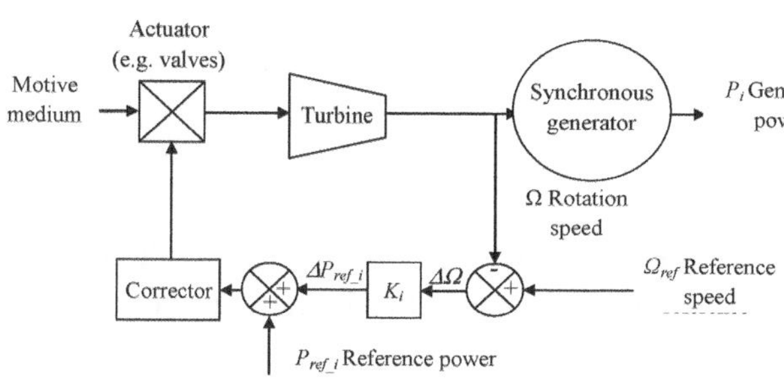

Figure 3.6. *Operating principle of primary frequency control [KAN 14]*

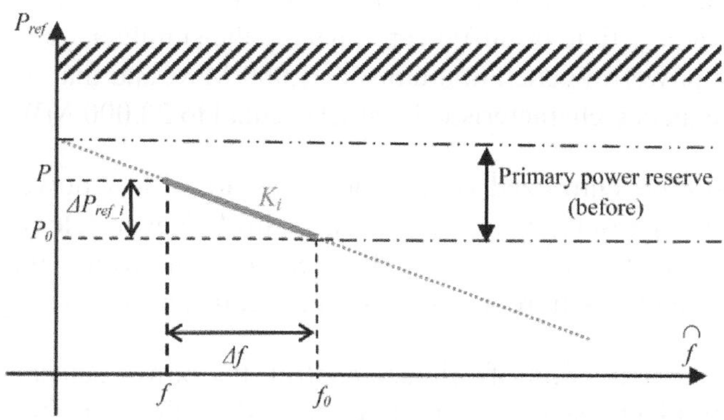

Figure 3.7. *Primary frequency control static characteristics (drop)*

The linear power/frequency characteristics of primary control, also called "droop", can be written as:

$$P - P_0 = K_i \cdot (f - f_0) = K_i \cdot \Delta f = \Delta P_{ref\_i} \qquad [3.3]$$

where f_0 and f are the reference frequency and the grid frequency, respectively, K_i is the power frequency of this generation unit i and P_0 and P are the reference power and its instantaneous power, respectively. A similar equation can be written for each generation unit participating in primary control at a given time. Thus, their sum gives, on the scale of an electrical system containing n' machines contributing to primary frequency control:

$$\Delta P_{res} = K \cdot \Delta f \qquad [3.4]$$

with $K = \sum_{i=1}^{n'} K_i$ [3.5]

where ΔP_{res} is the primary reserve released when the frequency variation stabilizes at value Δf and K is the network power-frequency characteristic.

In Europe, to achieve the given tasks in terms of frequency control performance, the European Network of Transmission System Operators for Electricity (ENTSO-E) recommends that the primary reserve available at any time be sufficient to handle the simultaneous loss of the two most powerful generation units, or 3,000 MW, and that the network power/frequency characteristic be at least equal to 20,000 MW/Hz.

Because decentralized generation assets are of low power, they do not usually participate in frequency control. Nevertheless, the grid codes define the conditions of their connection, which include, for example, certain requirements regarding frequency ride-through.

After the activation of primary control, the power set points of the generation plants have been modified and, since the correction made is proportional, a static frequency deviation remains: the frequency stabilizes at a different value of f_0 (see Figure 3.5). Moreover, all the generation plants in an interconnected grid make their contribution no matter the source of the disequilibrium, which impacts commercial exchanges at the boundaries, rendering a second level of control necessary.

3.2.4.1.2. Secondary frequency control

Secondary control was introduced in order to re-establish international exchanges and cancel static frequency error following primary frequency control actions. It is automatic and can be one of three types, according to [ENT 14]:

– centralized, that is, done for a countrywide grid by a single controller (as is the case in France);

– pluralist, that is, in a country divided into several electrical zones, with each zone controlled independently (as is the case in Germany);

– hierarchical, which corresponds to a pluralist type with a centralized controller coordinating the various zones (as is the case in Spain).

To fulfill the tasks stated above, secondary frequency control must be carried out in order to maintain the control error G at nearly zero:

$$G = \Delta P_i + \lambda \cdot \Delta f \tag{3.6}$$

with $\Delta P_i = P_{exch} - P_{prog}$ [3.7]

and $\Delta f = f - f_0$ [3.8]

where P_{exch} is the total amount of power exchanges on interconnection lines, P_{prog} is the scheduled exchange of power, ΔP_i is the deviation on exchanges, λ is a coefficient called the secondary power/frequency characteristic (MW/Hz) and Δf is the frequency deviation from its set-point f_0.

In France, a centralized controller located at the national dispatching center remotely modifies the operating point of generation plants participating in secondary frequency control in order to cancel out the error G. To do this, the controller calculates a control set-point $N(t)$, falling between −1 and 1, and transmits it to the generation units contributing to secondary control [RTE 14]. The term G/λ is called the Area Control Error (ACE) and the level $N(t)$ is calculated by:

$$N(t) = -\frac{\alpha}{P_r} \cdot \int_{-1}^{+1} \left(\frac{G}{\lambda} \right) \cdot dt + \frac{\beta}{P_r} \cdot \frac{G}{\lambda} \qquad [3.9]$$

where α is the control slope (MW/turn), β is the proportional gain (MW/Hz) and P_r corresponds to the secondary power reserve of the control area (MW). P_r is the sum of the secondary power reserves (p_r) of each of the generation plants participating in secondary control. As an example, for a nuclear plant of 900 MW contributing to secondary frequency control at a given time, p_r is equal to 5% of the nominal power, or 45 MW. The secondary power reserve recommended by ENTSO-E in each control area depends on its maximum consumption during the day according to the curve in Figure 3.8:

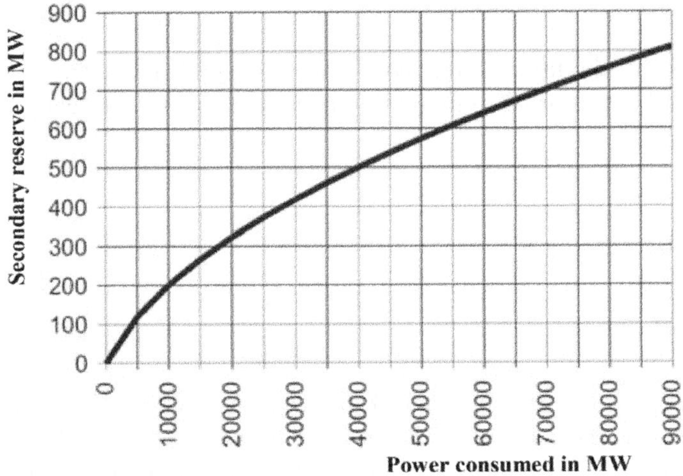

Figure 3.8. *Recommended secondary reserve volume [ENT 14]*

Level $N(t)$ is sent to the generation plants in order to control their power set-point. To limit stress on hardware, the variation slope of the signal $N(t)$ is limited. Thus, in normal operation, the shift from $N(t) = -1$ to $N(t) = +1$ takes place in 13 min, which corresponds to 7 nw per minute for a nuclear plant of 900. In the event of a large control error G, the shift from $N(t) = -1$ to $N(t) = +1$ can take place in 2 min.

3.2.4.1.3. Tertiary frequency control

In the event of major incidents, primary and secondary power reserves can be partly used or even exhausted, and it is then necessary to rebuild them; this is the objective of tertiary control, which is manual and takes place during a third stage. Calling upon generation owners and some industrial consumers to modify their operating plan, it is used to "rebuild primary and secondary power reserves, to compensate for profound and lasting imbalances between supply and demand and to maintain operating margins at adequate levels" [HAR 05].

To carry out this control within a liberalized electricity market, the issue was to define a means that would allow clear, non-discriminatory activation of participants while satisfactorily fulfilling system security imperatives. One mechanism, called a "balancing mechanism", has been implemented and is coordinated by RTE from the national dispatching center. This is a permanent call for bids enabling the French transmission system operator to have real-time access to the power reserves necessary for system security and reliability. Adjustment offers, whether they are "upward" (increased production, increased importation, reduced exportation or reduced consumption) or "downward" (the opposite of the former) are proposed by stakeholders who have signed an agreement to participate in the balancing mechanism: producers, foreign stakeholders operating through interconnections, industrial customers, etc. If necessary, the transmission system operator can activate at any time the offers meeting his needs, in an order defined according to the principle of economic precedence. Producers connected to the transmission grid operated by RTE are obligated by the French regulations to offer all available generation capacities to the adjustment mechanism. In addition, contractual agreements are established upstream of real time in order to guarantee the availability of tertiary reserve volumes that can be provided in less than 15 min and in less than 30 min.

3.2.4.2. *Guarding against voltage collapse*

Given its impact on the operation and lifespan of electrical facilities, the maintenance of the voltage level within limited ranges of variation is a vital criterion in the assessment of the quality of the electrical energy supply. The range of variation of +/–10% on either side of the nominal rms value is standardized in Europe by standard EN50160 [EN 99]. For the transmission and distribution grids in France, the variation ranges of voltage are given for illustrative purposes in Table 3.2.

	Voltage level (kV)	Normal operation		Emergency operation	
		Min (kV)	Max (kV)	Min (kV)	Max (kV)
Transmission grids	400	380	420	320	440
	225	200	245	180	250
Distribution grids	90	78	100	72	102
	63	55	72	50	73.5

Table 3.2. *Ranges of voltage variations in the French transmission and distribution grids [RTE 14]*

Unlike frequency, which is a global system state variable (in steady state) resulting from the instantaneous balance between supply and demand, voltage is mostly a local quantity. Its complete control sequence is described in [RTE 04] and [RIC 06], from which we have taken Figures 3.9 and 3.10, respectively. Because the transmission and distribution grids are mainly inductive, voltage drops in HV networks are mainly due to the circulation of reactive power. Therefore, it is via the control of this quantity that voltage control is carried out on the transmission system.

Voltage control is hierarchized in time and space and is based on the coordination of:

– a rough, slow compensation, as closely as possible to the loads, of the reactive power they draw. Besides a billing of the reactive energy applicable in certain conditions to industrial customers (incentive to install local compensation), reactive power compensation is mostly carried out in distribution systems via capacitor banks installed downstream of HV/HV transformers in primary substations. The switching on and off of these capacitor sets is controlled automatically by varmeter relays, thus making it possible to maintain exchanges at the transmission–distribution interface beneath contractual thresholds;

– the precise and dynamic action of the centralized generation plants that establish the overall reference voltage for the whole system at the transmission level grids. As with frequency, voltage control is broken down into three levels as illustrated in Figure 3.10; more details are provided in the following sections.

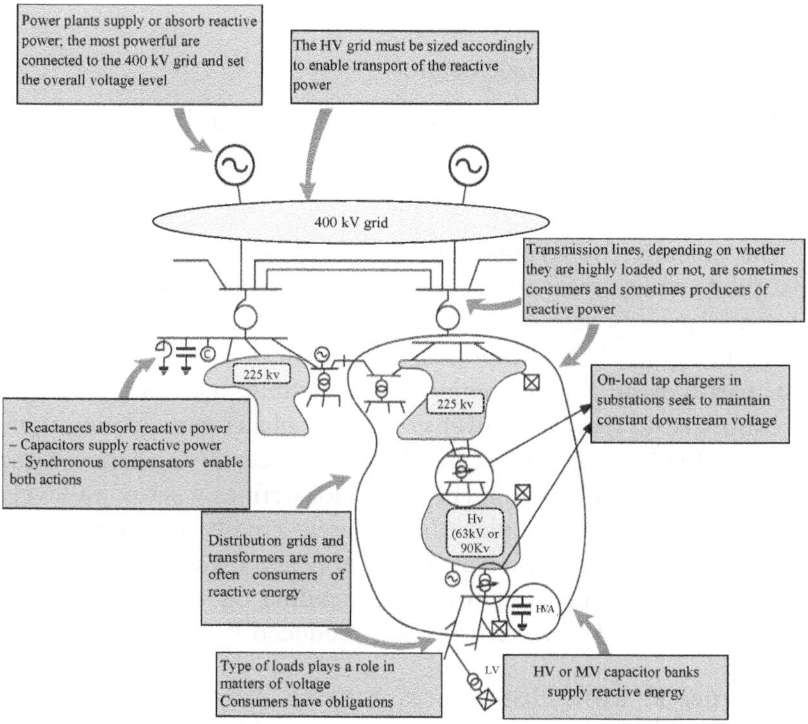

Figure 3.9. *Positioning of various pieces of equipment contributing to voltage control [RTE 04]*

In addition to limiting voltage drops in HV grids, local compensation is important, because (1) it limits apparent currents and thus losses on the grid and (2) it makes it possible to preserve the availability of reactive power reserves on centralized generation assets for precise/dynamic control and response to incidents.

3.2.4.2.1. Primary voltage control

Primary voltage control is automatic and is carried out in high-power generation plants connected to the transmission grid. It keeps the output voltage V_g of each alternator at the reference power level ($V_g^{ref} + \Delta V$) where V_g^{ref} is the reference voltage of the generator and ΔV is the correction defined by secondary control. The time constant of primary control is usually in the order of a few hundred milliseconds.

3.2.4.2.2. Secondary voltage control

As part of secondary control, the electric grid is divided into different control areas, each having a pilot node, as shown in Figure 3.11.

The objective of secondary control is to control the voltage V_p of this pilot node, selected in area for its representativeness, at a reference quantity determined by the transmission system operator. As shown in Figure 3.10, the process is divided into two parts, with the first part centralized at regional dispatching and the other part decentralized to the high-power generation plants participating in this control. More specifically, at the dispatching center in charge of each zone, a proportional integral (PI) corrector defines a set-point level N to send to the generation plants in the area. Finally, the decentralized part of the mechanism receives this level N and uses it to calculate the contribution of its generation plant depending on various parameters including voltage V_g and active power produced P_g.

In order to avoid disturbances between primary and secondary control, the response time of secondary control is slower, in the order of one minute.

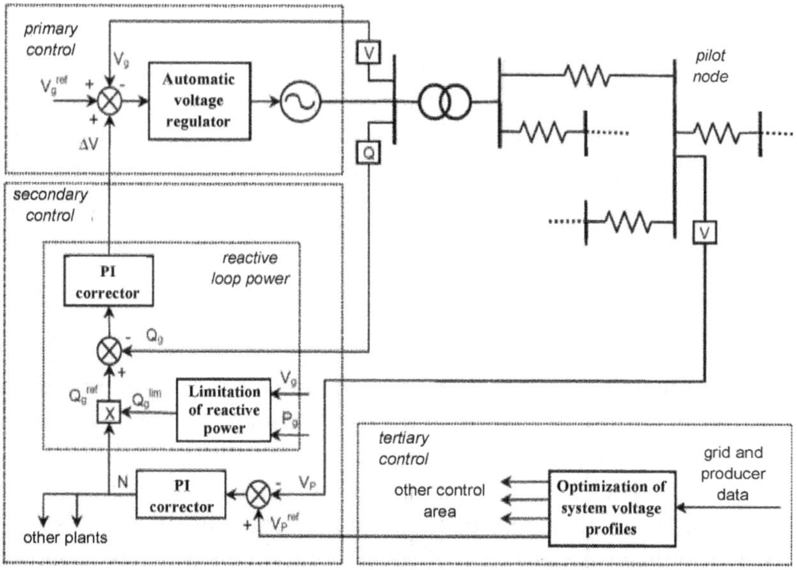

Figure 3.10. *Hierarchy of voltage control [RIC 06]*

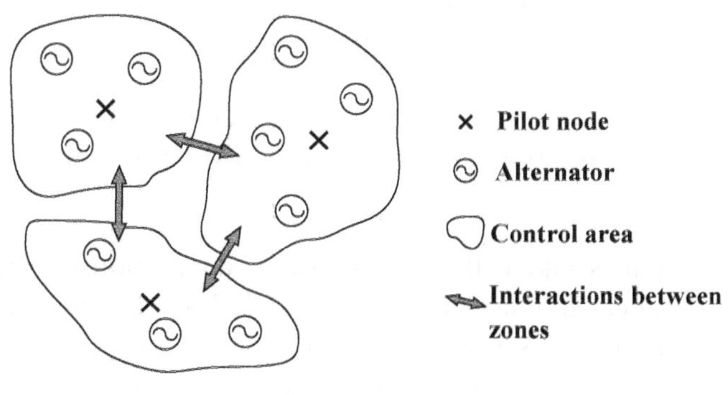

Figure 3.11. *Secondary voltage control [RIC 06]*

3.2.4.2.3. Tertiary voltage control

Tertiary control is centralized at the national dispatching center. It consists of updating the reference voltage of the pilot nodes every 15 min (V_p^{ref}). This operation is carried out using an optimal power flow (OPF) calculation code and is intended to guarantee good overall voltage maintenance and coordination between the reference power levels of the different zones, in order to prevent possible incompatible actions, for example.

3.2.4.3. *Guarding against losses of synchronism*

Synchronism is the fact that all of the power plants interconnected in the electric grid operate at the same frequency. However, after a short circuit on the grid, plants close to the site of the short circuit may undergo an acceleration in the rotation speed of their alternator [KUN 94]. If the alternator cannot resynchronize itself, there is a loss of synchronism. To avoid losses of synchronism, it is important to eliminate faults as quickly as possible and to have sufficiently robust voltage and speed control systems in operation [RTE 04].

3.2.4.4. *Guarding against cascading overloads*

In the power system, the distribution of power flow depends on the localization of generation, consumption, reactive power compensation equipment and grid impedances. However, flows must remain limited including in accordance to maximum current IMAP as defined in section 3.2.2. Congestion appears when the IMAP of a line or cable is exceeded.

If no action is taken, there is a risk of congestion damaging the conductor or, heated under the effect of Joule loss, the line may stretch out, thus reducing safety distances in relation to the ground (vegetation or constructions), which can initiate an electrical arc or compromise the safety of people or hardware. However, by design, IMAP overloads can be tolerated for a limited time, for example, in the case of a 15% overload, the time period is 20 min on a 400 kV grid and 1 min for 90 kV systems. Figure 3.12 illustrates the various trigger thresholds for 400 kV overhead lines [VER 09, VER 10, VER 11a].

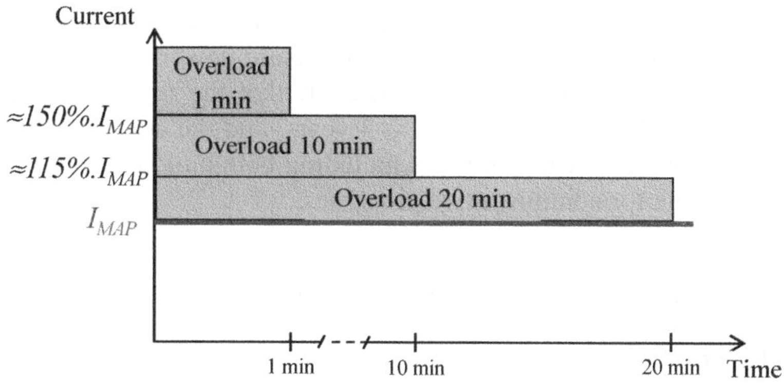

Figure 3.12. *Overload capabilities of 400 kV grids*

In a situation where the dispatcher has been unable to find a way to eliminate congestion in a line within the allowable time, the line is opened automatically using overload relays. The power that was flowing through this the line is then transmitted to other assets still in operation, which may then be overloaded as a result. If corrective decisions are not quickly taken, other outages and subsequent load transfers will appear; this is the phenomenon of cascading overload. This phenomenon has often resulted in widespread incidents such as those of 19 December 1978 in France [RTE 04], 28 September 2003 in Italy [UCT 04] and 4 November 2006 on the UCTE grid [UCT 07].

Consequently, congestion management is of vital importance, not only for the safety of people and goods, but also for power system security and reliability.

Currently, the number of congestions may increase due to decentralized generation which, by injecting massive amounts of power into distribution grids, can cause an inversion of power flows. This phenomenon is already appearing in the interconnection lines between Germany and its neighboring countries due to the high concentration of wind energy in the northern part of the country [EON 05].

In order to prevent the appearance of congestion events on constrained transmission lines and thus ensure power system security, new lines must be built. Note, however, that the time necessary to reinforce a substation can be up to 5 years, and the construction time for a new line can reach 10 years including both building time and the required time for administrative processes.

3.3. Services that can be provided by storage

3.3.1. *Introduction*

An electrical energy storage system can be defined very generally as an asset with three capacities: (1) to consume electricity, (2) to accumulate the corresponding energy and then (3) to use this energy to produce electricity [RUE 12]. These three capacities are interdependent, which requires suitable control strategies. For example, storage can produce power only after having consumed it; similarly, since the maximum amount of energy that can be accumulated is physically limited, the ability to consume power may be lessened or even unavailable if the generation (discharge) phases are not carried out.

From the operators of conventional or renewable generation assets to the final consumers and including the transmission and distribution system operators, storage can provide services to various stakeholders in power systems.

These services are based on the potential functions of storage such as (1) the planned displacement of energy blocks in time, (2) the building of a power reserve that can be used to serve power system needs to support critical usages in the event of an outage, (3) the provision or absorption of reactive power or (4) the active filtering of disturbances affecting the voltage wave (harmonics, imbalances, etc.). The two latter examples, (3) and (4), do not use the actual "energy accumulator", but rely on capacities that may be offered by the equipment necessary to interface it with the grid, namely rotating machines or power electronics converters. Therefore, they cannot be the only reason for a decision to invest in a storage system, but may help to make such a device profitable by adding to the revenue generated by the operation of the energy accumulator itself. For a given storage function, the need(s) satisfied and the process(es) of value creation can be very different depending on the stakeholder considered. For example, by making it possible to adjust the load curve, the transfer of energy blocks noted in (1) above can make it possible:

– for the operator of a mix of generation assets to limit the use of costly peaking plants;

– for a grid operator to relieve overload on constrained assets whenever necessary and thus to postpone or even avoid reinforcement decisions;

– for a final consumer to reduce his bill by taking maximum advantage of the rate offered.

This section is intended to give an overview of the various services that can be provided by energy storage in power systems. In each case, we will attempt to clarify the stakeholder involved, the operating principles, technical requirements and when possible, value making.

3.3.2. *Services required for connection to the transmission grid*

Above certain power thresholds, electricity generation plants connected to the transmission grid are required to participate in its operation. Bulk storage units are therefore concerned; smaller systems may also possibly offer their contribution in a context where the provision of these controls is opened up to a higher number of stakeholders. Required services are defined by the transmission system operator in order to ensure system security in the face of various uncertainties and hazards. These services are collectively referred to in the literature as "ancillary services".

3.3.2.1. *Contribution to frequency control*

The principle of frequency control has been described in section 3.2.4.1. In the absence of rules specific to storage at present, Table 3.3 shows the requirements to be satisfied in France by conventional generation units.

As part of primary frequency control, any generation plants of more than 40 MW must be capable of providing a power reserve of 2.5% of its rated power. In addition, the plant must be able to maintain the activation of the reserve for 15 min at least, and the power reserve must be totally released (1) for a frequency variation not exceeding 200 mHz and (2) in less than 30 s. In 2014, the payment for primary control in France was approximately €9/MW per half-hour of provision.

Service	Minimum reserve	Duration of delivery	Maximum response time
Primary frequency control	$P_{inst} \geq 40$ MW: $P_{rés} \geq 2.5\%$ of P_{max}	≥15 min	Totality in ≤30 s Half in ≤15 s
Secondary frequency control	$P_{inst} \geq 120$ MW: $P_{rés} \geq 4.5\%$ of P_{max}	holded as long as necessary	2–13 min between $N = -1$ and $N = 1$

Table 3.3. *Technical data for frequency control services [RTE 14]*

With regard to secondary control, power plants with nominal power higher than 120 MW must be able to provide a power reserve of at least 4.5% of their nominal power. The payment for secondary frequency control in France is composed of two terms. The first corresponds to the availability of the reserve and was approximately €9/MW per half-hour in 2014, while the second corresponds to the use of the reserve and is equal to approximately €10.5/MWh. It is paid by RTE to the producer in the event of an increase in generation, and the reverse is true in the event of a reduction in generation. It is worth noting that a second term similar to that of secondary control was recently added to the rules related to primary control (same amount in €/MWh as secondary control).

Currently, the provision of these services tends to be increasingly open, giving conventional electricity producers the possibility of meeting their requirements by relying on the contribution of other stakeholder in the electrical system. Thus, the emergence of markets where prices will be freely negotiated is tending to occur and will be discussed in section 3.3.6. In any case, as mentioned in section 3.2.4.1, the total required power is limited, for example, for primary control, the total volume is 3,000 MW in Europe and on the order of 600 MW in France.

3.3.2.2. *Contribution to voltage control*

The principle of voltage control, based mainly on controlling reactive power, was described in section 3.2.4.2. Table 3.4 shows the technical data to be satisfied in France by generation units connected to the transmission grid.

Contribution to voltage control is made via the consumption or production of reactive power, which impacts the sizing of the energy storage system only slightly (increase in the ratings of the grid interface but no impact on the energy capacity). For example, in the context of pumped hydro storage or compressed air energy storage (CAES), the alternator(s) must be adapted for this service, which is limited by physical constraints: maximum stator intensity, maximum rotor intensity, maximum internal angle, minimal and maximal stator voltages in steady state, etc. In the case of sources or storage systems connected to the grid via power electronic converters, the latter are used to control reactive power at the point of connection and must therefore be sized for the apparent power exchanged. For voltage source converters, limits exist that are dependent on the real power generated, the continuous-side voltage of the converters and the inductance of the electric grid connection filter [BOU 09, DEL 10].

Service	Minimal range	Duration of delivery	Maximum response time
Primary voltage control	Primary: plants connected to HV Secondary: plants connected to HV-2/3 *U*/*Q* diagrams for different power levels; the value is a Q/P_{max} ratio on the order of ±0.3	Continuous	A few hundred milliseconds
Secondary voltage control		Continuous	In the worst-case scenario, the response dynamic must be better than that of a first order with a time constant on the order of approximately 60 s

Table 3.4. *Technical data for voltage control services [RTE 14]*

Payment for these services in France depends on the location, as certain areas of the electric grid are more or less sensitive with regard to reactive power. In France, the payment for primary voltage control in sensitive areas was, in 2014, composed of a fixed part of €761/MVA/year and a variable part on the order of €0.03/Mvar per half-hour of operation. The remuneration in normal areas corresponds to the variable part only. Payment for secondary voltage control is an increase of 50% in the variable part of primary voltage control.

3.3.3. *Potential additional services provided to a transmission system operator*

As well as the above-mentioned services that are mandatory to require a connection to HV grids, stakeholders in the power system may provide additional services to help the transport system operator ensure system security and reliability.

3.3.3.1. *Tertiary frequency control reserve*

The principle of tertiary reserve was described in section 3.2.4.1. There are several tertiary reserves that are used in case of need, and differentiated according to their mobilization time. Thus, the so-called "fast" tertiary reserve corresponds to a reserve that can be provided in less than 15 min, while for a response time of between 15 and 30 min, an "additional" tertiary reserve is used. In France, the volumes are approximately 1000 MW for the fast tertiary reserve (which can actually be provided in less than 13 min) and approximately 500 MW for the additional tertiary reserve, respectively. Table 3.5 gives some data concerning tertiary reserves in France.

Service	Minimum total volume	Maximum response time
Fast tertiary reserve	Approximately 1,000 MW in France	15 min
Additional tertiary reserve	Approximately 500 MW in France	30 min

Table 3.5. *Some technical data concerning tertiary reserves*

In order to ensure the availability of fast and additional tertiary reserves, a contract is established between the transmission system operator and the stakeholders who agree against a fixed premium to make this fast-accessible power reserve available to the balancing mechanism. In the case of a storage system, the requirement of being able to release the reserve for a period of several hours may require substantial sizing in terms of energy. Moreover, the fact of guaranteeing the availability of this service greatly limits the possibility of using storage for other purposes.

3.3.3.2. *Congestion management*

Congestion management consists of relieving a transport line temporarily overloaded beyond the maximum permissible by the flow of power. Due to their reversibility, storage systems are of interest as means to resolve congestions. Figure 3.13(b) illustrates the principle of storage for treating a congestion event that corresponds here to an overload to 125% of the line between 2:00 and 5:00 due to excessive load consumption (Figure 3.13(a)). Storage injects power during this period in order to relieve the line and is recharged when the flow in the line is minimal.

However, the means for treating congestions are chosen by the transmission system operator from among the offers made on the balancing mechanism (see section 3.2.4). To participate in this in France, in current conditions, a storage system must be able to make offers of at least 10 MW.

Congestions that appear frequently are a sign that the grid infrastructure has reached the limits of its normal operation. In a case such as this, the transmission system operator must eventually reinforce the grid by constructing additional electrical lines; however, the construction cost is high, and the process may take more than 10 years. In this context, the implementation of a storage system may be an appropriate alternative [VER 09, VER 10, VER 11a].

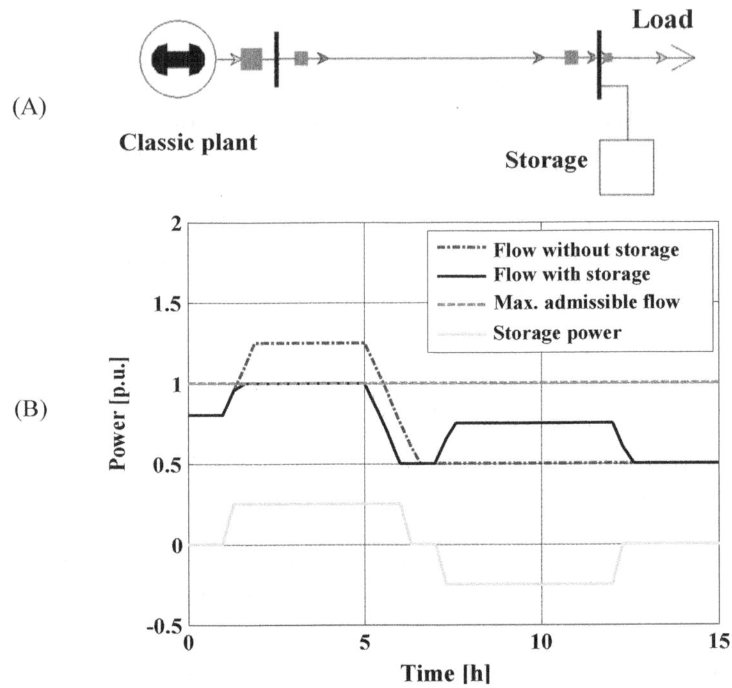

Figure 3.13. *Principle of using storage to manage congestion. For a color version of the figure, see www.iste.co.uk/robyns/powergrids.zip*

3.3.3.3. *Black start*

Subsequent to exceptional events leading to a collapse of the electrical system, or a blackout, the transmission system operator must re-energize the grid in order to resume the power supply to consumers. To do this, he uses generation plants that can be started up without requiring a grid (black start) or plants that have successfully been kept online during the incident (islanding). Energy storage devices can contribute to this grid restoration; however, they must fulfill the following requirements [RTE 14]:

– control the transient created by the switching-on of load blocks: the storage system must be able to release a power level of at least 5–10% of the net continuous power upon request in stages;

– stabilize the separated grid whatever the power level requested by the loads: the storage system must be able to operate across a wide power range, in particular at low power levels, and to maintain operation at these levels;

– preserve the overall balance between supply and demand while remaining within an acceptable frequency band: the system must be able, upon request from the transmission system operator, to provide a frequency reference to the isolated grid as long as the it has access to adequate power reserves;

– maintain stable operation without sustained frequency or voltage oscillations: the system must have appropriate speed and voltage controllers;

– control overvoltages linked to recoupling with any other part of the grid otherwise reenergized;

– tolerate up to 12 power inversions without damage during coupling operations between regional feeders.

3.3.4. *Potential services provided by storage to a distribution system operator*

3.3.4.1. *Peak shaving*

3.3.4.1.1. Principle

Peak shaving in order to delay or even avoid expansion on distribution grids is a service often imagined for decentralized storage. It is notably described in publications such as [EPR 02, EPR 03, EYE 04, EYE 05] and [EYE 10].

When the nominal capacity of a piece of grid equipment such as a line or a transformer risks being exceeded in the event of a load increase to be powered and/or generation to be evacuated, the conventional solution implemented by distribution system operators consists of either a reinforcement of existing infrastructures or the construction of new ones. Due to the standardization of hardware, the literature emphasizes that the capacity increase thus achieved is often much greater than the short-term needs, which leads to the "underexploitation" of the new assets for many years – in other words, to overspending with regard to immediate needs.

Peak shaving consists of using a storage unit as a temporary (at least) solution to relieve pieces of grid equipment that have reached their limits. In the case of constraints related to upstream-downstream power flows (load), as shown in Figure 3.14, storage is charged when the power flow is low in order to have access to a reserve of energy to be released at peak time, thus minimizing extreme currents traveling through the pieces of grid equipment concerned. In the case of constraints related to downstream-upstream power flows (local generation), discharge is carried out when the grid capacity enables it, in order to store power during peak production times. Whatever the origin of the constraints, these actions on active power profiles can be completed by controlling reactive power as well. For example, on a given piece of grid equipment, bringing the tangent *phi* of an initial value $\tan(\varphi) = Q/P = 0.4$ to a final value $\tan(\varphi) = 0$ will reduce apparent flow by 7%.

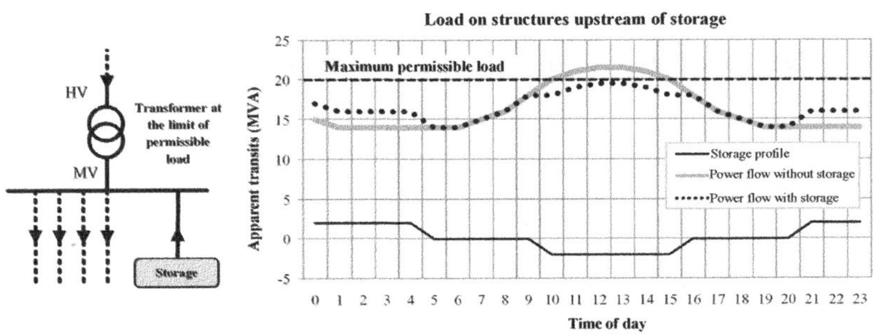

Figure 3.14. *Peak shaving using local storage: illustration in the case of a load peak in a power substation transformer [DEL 09]*

Thus, temporarily shaving load on generation peaks using storage makes it possible to postpone or even avoid expansion on distribution grid infrastructures. A practical example of this is given by [NOU 07], which details the use of a 1 MW/7.2 MWh sodium–sulfur battery on the 12 kV grid of the American utility company American Electric Power (AEP) in 2006. This storage facility made it possible to extend the use of a 20 MVA transformer for several years, thus delaying estimated expenditures of $2 million. More recently, UK Power Networks has experimented with a 600 kVA/200 kWh system based on a lithium-ion battery, and in [UK 14] it confirms the feasibility of limiting power peaks provided that the demand curve can be accurately forecasted.

At the end of the investment deferral period, the necessary reinforcements are carried out; the capacities of the structures added are then better exploited than if they had been built several years earlier. If it has not reached the end of its lifespan, the storage system can be either kept in place to be used in a new service offer, or moved to another node in the grid if its design enables it.

This service may prove to be of interest in certain specific cases, in particular when a constraint (environmental, legal, etc.) hinders or delays grid reinforcement, posing the risk of harming the quality of electricity supply, or in order to avoid costly grid infrastructure developments for a temporary need.

In the case of AEP described above, a risk of constraint in normal operations was the original reason for the implementation of a storage system and the eventual upgrading of the substation, which had reached its limits. That being said, section 3.3.4.1 could also apply to situations where reinforcement would be required according to the criteria used for sizing the emergency configuration(s) planned in the event that a part of a distribution system becomes unavailable. These types of configurations in which "N-1" requirements are most problematic in terms of planning may lead to an oversizing of equipment in relation to the needs of normal operation, as rare events are planned for. Similarly, for users, incorporating peak shaving into the planning of contingency grid schemes may limit the necessary amount of investment.

Finally, note that the service we have just described for a distribution system operator may also be of interest to other stakeholders who would have to finance all or part of a reinforcement project or a new construction (line, transformer, etc.); this is currently the case in France for a user who requests the connection of a new facility. The situation of a consumer will not be addressed in this chapter, but that of a decentralized generator is discussed more broadly in section 3.3.5.5.

3.3.4.1.2. Technical requirements

With regard to the connection point, investment deferral requires us to set up a storage system in order to be able to act on stressed infrastructures, which can leave more or less flexibility depending on the case. When free choice is possible, the parameters to be taken into account must be studied on a case-by-case basis: availability of land, accessibility, acceptability, possibility of combining multiple services, communication needs, etc. Sizing in terms of power depends on the load at the time of the study, its growth rate and the deferral time desired. For example, an increase in the load of 2% per year requires a minimum, respectively, of 20 kW (1 year), 104 kW (5 years) and 220 kW (10 years) for each megawatt of initial peak power flow on an infrastructure. Under a realistic hypothesis (several years of postponement and moderate load growth), the power required for peak shaving falls between 500 kW and several megawatts for an HV plant against several hundred kilowatts at LV [EPR 03, EYE 05, IAN 05]. The necessary discharge time depends on the form of the load profile and can be estimated at between 2 and 10 h. Conversely, there is no requirement with regard to storage response time, which can exceed several minutes given the temporary overload capacity of distribution grid equipment.

3.3.4.1.3. Value making

When there is a risk that technical limits will be reached, the decision of a grid operator to undertake works is not bound by economic justification. However, if several solutions are possible (reinforcement of existing assets, creation of a new feeder), the choice will fall on the one that minimizes the latest assessment of costs over the lifespan of the project, taking into account investment costs, maintenance coasts, purchase of electricity losses and impact of breakdowns [ERD 08, DOU 02].

The economic interest of peak shaving must therefore be studied in this context, comparing the “storage” option to the other options available. This is why the literature generally defines the value of this service as the difference between the net present cost with the deferral and the conventional investment cost. The results obtained are dependent on the conditions of each project, with the service being all the more interesting when the works considered are capital-intensive. Numerical applications on real data are notably offered by [EYE 04, IAN 05] and [NOU 07], with results ranging from a value of 0 to $1.5 million per MW of storage installed for the most extreme cases.

One question remains, however, without reply: in the hypothesis that an attractive situation was identified, how would the value created in terms of expenditure postponement be translated into remuneration for the storage asset? The answer comes naturally if the operator of this storage system is the grid operator himself, as this is the case with a vertically integrated utility such as AEP in the example cited above. On the other hand, if the operator is a third party, the mechanisms allowing a distribution system operator to make requests and give compensation in a clear and non-discriminatory manner for local flexibilities such as decentralized storage are yet to be determined in France.

3.3.4.2. *Local voltage control*

3.3.4.2.1. Principle

Maintaining the value of the supplied voltage is needed to enable the normal operation of electrical appliances and to guarantee their lifespan. This is why an admissible range for the 10 min averaged rms voltage is set by regulations in France. For example, the interval is ±10% on either side of the nominal voltage for any MV or LV user (decree 2007–1826 and its ordinance of 24 December 2007). This minimal requirement is given, sometimes in a stricter form, in contracts offered by distribution system operators.

Due to line and cable impedance, power flows cause drops in voltage, often quantified via the approximation:

$$\frac{\Delta U}{U} \approx \frac{RP + XQ}{U^2} \quad [3.10]$$

where U is the phase to phase voltage, R and X are the resistance and reactance, respectively, of the section considered and P and Q are the active and reactive power flows, respectively.

In France, distribution system operators have access to different resources to meet regulatory or contractual requirements regarding voltage level, including the on-load tap changer with which HV/MV transformers at power substations are equipped. As described in [RIC 06], for example, this mechanism controls voltage at the level of the MV bus bar as close to a reference value as possible, by continuously adjusting the transformation ratio via timed tap changes. Without decentralized generation, voltage decreases from the primary substation to the end of the feeder (see gray curves in Figure 3.15); in this case, the bus bar reference is set near the upper limit in order to avoid low voltage constraints at the end of the feeders at peak load. The connection of a producer reduces or even reverses power flows, and thus tends to increase the voltage, which can prove limiting, particularly when consumption is low (see the black curves in Figure 3.15).

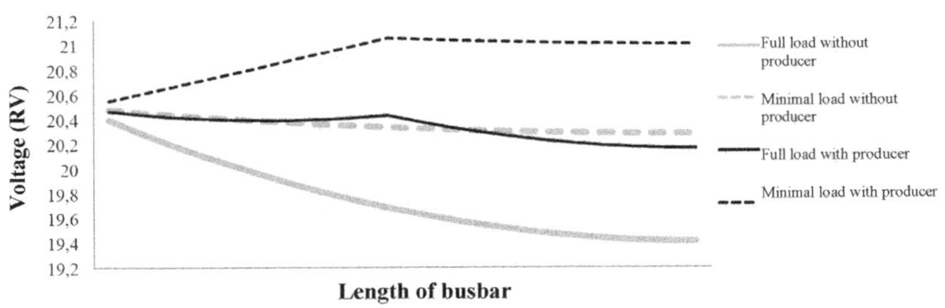

Figure 3.15. *Voltage plan along an a MV feeder with and without a distributed generation [DEL 09]*

Grid operators' planning studies are conducted in order to maintain a satisfactory voltage level to users. If a constraint is identified, classic options consist of the adaptation of the feeder concerned (increasing the section of the most high-impedance conductors) and can go so far as to include the creation of a dedicated feeder, for example, for the connection of the most powerful decentralized generation assets. When several options are technically possible, the one that minimizes the net present cost over the lifetime of the project (see section 3.3.4.1) will be chosen in the end [ERD 08].

This potential service consists of controlling the injections of real power and possibly of reactive power of a storage system on a constrained feeder in order to ensure adequate voltage during peak load times (prevention of low voltage situations) or peak generation times (prevention of high voltage situations). In such cases, it is an alternative to investments that would have to be made otherwise to maintain the quality of supply within permissible contractual or regulatory limits.

For example, a 350 kVA/8 h Vanadium redox-flow battery commissioned in 2003 for the American utility company PacifiCorp is presented in [EPR 05a]. The system, installed in the middle of a 25 kV feeder that was very long and difficult to reinforce in a protected environment, ensured maintenance of the voltage level for several years via a preplanned charge/discharge profile. Technically, it is possible to equate this service with the peak shaving described earlier in section 3.3.4.1, with the main difference having to do with the limits, the attainment of which the storage system serves to postpone (currents vs. voltage).

Local voltage control was also tested by UK Power Networks with the 600 kVA/200 kWh system connected at 11 kV and tested between 2011 and 2014. The report [UKP 14] presents the results, which effectively show a reduction in local voltage variations in response to controls applied at the terminals of the storage plant on real and reactive power, with the latter also being more effective on the test grid.

Like peak shaving, this service is interesting for stakeholders bearing the costs of developing the distribution grid. Aside from the operator itself, it may also be of interest for a user requesting the connection of a new facility to the grid. Consumer situations are not discussed in this chapter, but the situation of distributed generation is discussed more broadly in section 3.2.3.

3.3.4.2.2. Technical requirements

Since voltage is controlled at the primary substation by the on-load tap changer, its control by a storage system is possible only at a certain electrical distance from the substation, and is more effective when the impedance between it and the connection point is high.

Unlike peak shaving, which requires the insertion of the storage system "downstream" of constrained pieces of grid equipment in order to be able to act on the power that flows through them, this service gives more freedom. Weakly supplied customers at the end of a bus bar may, for example, benefit from support to the voltage profile provided upstream, in the middle of the feeder. This flexibility can facilitate the mutualization of this control with other storage applications, but requires the implementation of a communication network or estimation techniques adapted in order to be able to remove constraints from a remote site.

Using equation [3.10] with typical impedance values of HV and LV grids in France, power sizing can be proposed: between 100 kW (kvar) and several megawatts (Mvar) are required to have a significant impact on the HV voltage profile, compared to some 10–100 kW (kvar) in LV. If local support to the voltage is rendered using real power, the plant must be capable of executing a peak shaving comparable to that described in section 3.3.4.1, for a discharge time ranging from 2 to 10 h. However, since R and X have the same value in distribution grids, the best action to control the voltage may be real power, reactive power or a pairing of the two. Various storage supervision strategies can thus be defined, for example, in order to limit the energy sizing by giving priority to reactive power [KAS 07], an approach whose testing by UK Power Networks has confirmed its interest [UKP 14]. Finally, a response time of several minutes is enough to maintain the desired voltage level at an rms value averaged over 10 min.

3.3.4.2.3. Value making

The contribution of storage to local voltage control is value-enhanced via the postponement or avoidance of investments which, in the sense of conventional uses of grid development, would be necessary to maintain the voltage level within its contractual and/or regulatory limits. Methods used in the literature to quantify the benefits of investment deferral were described in section 3.3.4.1.

3.3.4.3. *Back-up power through intentional islanding*

3.3.4.3.1. Principle

With the increasing development of generation within distribution grids, the possibility of using these resources in emergencies in order to enable parts of the network to continue to operate during part or all of a local or generalized incident in the power system is of interest. However, most decentralized sources, such as photovoltaic plants, are not designed to operate independently from the grid; they cannot usually start by themselves, and their variable production makes it impossible to balance the demand. With adapted controls enabling it to become a safe, local voltage source without the grid (grid-forming capability), energy storage can be used by a grid operator during controlled islanding, allowing powering local loads to be secured. At least two scenarios are possible:

– mobile storage units could be used intermittently as generators. However, these technologies are more cumbersome than typical thermal generators, which is problematic when trying to reach areas that are difficult to access, for example, after a storm;

– voltage restoration in local networks could be added to the set of services offered by a stationary storage system, for example, in an area where conventional security solutions such as meshing (with normally open points) or reinforcement are difficult to implement.

Figure 3.16 illustrates how a service like this can be given by a storage unit connected to a MV feeder bus bar including loads and generation units. In this scenario, operations are coordinated in real time by the grid operator:

– in A, the *initial situation*, the storage system is assigned to its usual services; it is synchronized with the grid and controlled as a current injector;

– in B, a *fault appears on the feeder* upstream of the storage system connection point;

– in C, protection relays at both the feeder and storage system levels trip off to ensure the safety of people and goods. Consequently, *the feeder is no longer powered*;

– in D, the distribution system operator acts on the switches in order to isolate the defective local network. The storage system is then reconnected and, controlled as a voltage source, *supplies power to the safe islanded local network*;

– in E, the intervention of the grid operator has enabled the *elimination of the fault* and the voltage restoration of the whole feeder. However, the local network powered by the storage system remains islanded;

– in F, *the islanded power network is resynchronized with the grid voltage in order to be recoupled with it without a disruption*. The storage unit is then able to resume normal operations, and shifts back and forth between being a voltage source and a current source.

For more details, interested readers can read [MOR 07], for example, which provides details about the provision of back-up power to an LV grid via local production and a storage system. In practice, intentional islanding using decentralized storage systems has been used by the utility AEP, for example, and Nourai and Kearns [NOU 10] describe several mechanisms based on 2 MW/7.2 h sodium–sulfur batteries connected at MV. Moreover, the French demonstrator NiceGrid has planned to use a lithium-ion battery storage system in 2015 for the intentional islanding of an MV/LV substation powering consumers and of photovoltaic generation plants.

3.3.4.3.2. Technical requirements

The setup of this service particularly requires assurance that the protection scheme is adapted (grounding of neutral MV in islanded system, operating of protection relays at low short-circuit power, etc.) and that the control strategies used allow constant compliance with contractual or regulatory requirements with regard to power quality.

The power of the storage unit depends on the size of the local network recovered: from 10 kW to 1 MVA in LV, and in MV, from several hundred kilowatts for a limited area to 10 MW for several feeders. The discharge time at nominal power that must be available depends on the desired back-up time, which can, for example, be estimated according to the typical duration of the power outages against which the service is intended to guard. In terms of response time, the technology used must be capable of responding instantaneously to variations in generation and consumption in the islanded electrical network.

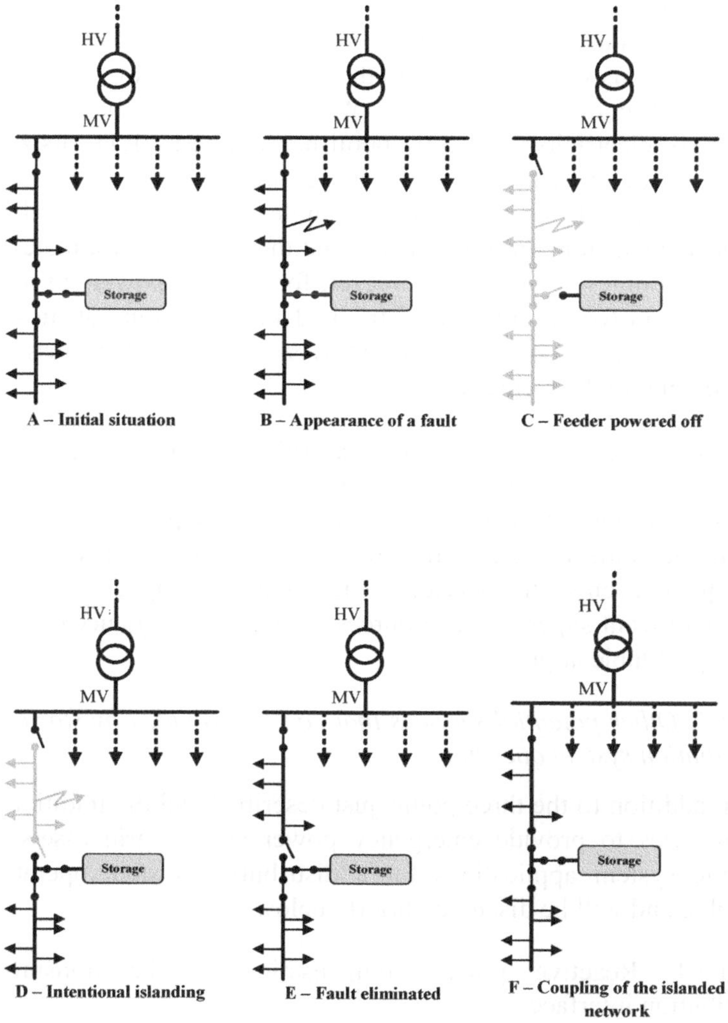

Figure 3.16. *Simplified illustration of the emergency back-up power provided to a part of the grid by a storage system [DEL 10]*

3.3.4.3.3. Value making

The value of local voltage control for the intentional islanding of local islanded networks can be approximated by calculating, on a case-by-case basis, the non-distributed energy (NDE) it makes it possible to avoid. This indicator, whose usefulness is described in [DOU 02], is used to estimate the cost of power cuts in order to make good decisions when investing in the reinforcement or development of the grid. According to the study "What value should be given to the quality of electricity?" recently published online by RTE, its average value can be currently estimated at €26/kWh, or approximately 200 times the cost of this electricity. A simple illustration that will give a better understanding of this concept and these types of estimates is the case of a domestic customer's freezer, for which a power cut of a few dozen hours (or several non-distributed kilowatt-hours at most) will cause the destruction of the food the freezer contains (for a loss of tens or even hundreds of euros).

Whatever the role NDE plays in a grid securing project, the cost of the "intentional islanding using storage" solution must be compared to that of the conventional or innovative options possible that would fulfill the same objective. In terms of the expenditure required for infrequent events, this service, which requires long discharge times and significant supervision resources, seems at first glance difficult to make profitable at present.

3.3.4.4. *Other potential services that can be provided by storage to a distribution system operator*

In addition to the three points just described and the traditional use of batteries to provide emergency power critical grid assets, other storage system applications for a distribution system operator are possible and will be discussed briefly below.

3.3.4.4.1. Reactive power compensation at the transmission–distribution interface

HV lines are mostly inductive ($X >> R$): according to equation [3.10], drops in voltage on the transmission grid are thus linked mostly to reactive power flows. Consequently, controlling voltage involves controlling reactive flows due to consumption by loads and by the grid (transformers and lines). Moreover, reactive power flows are the source of increased apparent currents and thus of Joule losses.

In this context, compensation is made in distribution grids via capacitor banks installed in HV/MV primary substations notably. Engagements and triggers of capacitor banks are controlled automatically by varometric relays [RIC 06]. In addition to the two reasons just mentioned, this supply as close as possible to the loads makes it possible to preserve the availability of reactive power reserves in centralized generation plants for precise/dynamic control (setting of a reference voltage profile on the major transmission grid) and responses to incidents [RTE 04].

A decentralized storage system could, via the possibilities offered by its power electronic converters, provide local reactive power compensation [EPR 02, EPR 05a]. Compared to the usual solutions, which are controlled in an "all or nothing" manner, this approach has the advantage of a more subtle management of reactive power injections. A service like this can create value via the deferral or even avoidance of investments on capacitor banks, or the cost of reactive power billed by the transmission system operator to distribution system operators according to the terms of TURPE in France. In version 4 of this rate, which became effective in 2014, the maximum amount is of the order of €15 per Mvarh supplied by the transmission grid (1) in excess of a contracted tangent *phi* and (2) at certain times of the year.

3.3.4.4.2. Reduction of Joule distribution losses

The electrical resistance of grid infrastructures induces losses proportional to the square of the current. Because it modifies power flows, storage has an impact on these, and adapted charge/discharge cycles might theoretically have a favorable energetic balance. The principle of this potential service is as follows:

– accumulation is completed during off-peak hours and increases currents on the distribution grid, resulting in an increase in the losses by $\Delta P_{J1} > 0$;

– discharge takes place at peak consumption times and the resulting current reduction reduces the losses by a value of $\Delta P_{J2} < 0$;

– the quadratic nature of the losses gives us $\Delta P_{J1} + \Delta P_{J2} < 0$ over a complete cycle, that is to say lower technical losses to be paid by the grid operator.

In order to characterize the possible interest of a "loss reduction" service provided by storage to a distribution system operator, case studies have been conducted on the simple model of a MV feeder presented in Figure 3.17. Assuming impedance and real and reactive consumption uniformly distributed over the whole length of the feeder, we get all of the calculations for the expression of instantaneous line losses $P_J(t)$ (see Appendix 1 of [DEL 10]):

$$P_J(t) \approx \frac{R}{U^2}\left[\left(P_S(t)^2 + Q_S(t)^2\right)x + \begin{pmatrix} P_S(t)P(t) \\ +Q_S(t)Q(t) \end{pmatrix}\left(2x - x^2\right) + \frac{P(t)^2 + Q(t)^2}{3}\right] \qquad [3.11]$$

where R is the total resistance of the feeder, U is the phase to phase voltage (assumed to be constant on the feeder), $P(t)$ and $Q(t)$ are the total real and reactive loads, x is the location of the storage system in the network ($0 \leq x \leq 1$) and $P_S(t)$ and $Q_S(t)$ are the real and reactive powers injected by the storage unit into the grid.

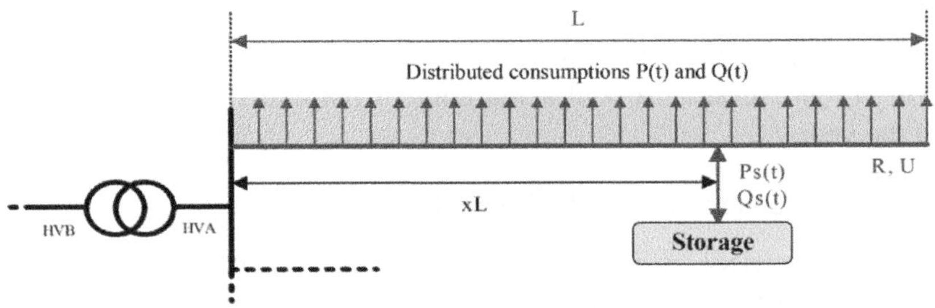

Figure 3.17. *Model of a MV feeder with uniformly distributed values*

This model has been applied to daily demand profiles and optimization calculations have been made to determine the location x and the power profiles $P_S(t)$ and $Q_S(t)$ that can minimize total losses.

The efficiency of the storage unit considered is 75%; the problem is constrained so that its state of charge is the same at the start and at the end of the day, which means that all of the charged energy is discharged, minus losses.

In these conditions, a detailed study over a realistic range of variation of input variables shows that storage effectively reduces online losses when its injections are carefully controlled, but that it always losses more energy than it makes it possible to save. As an illustration, on an MV feeder whose daily losses are approximately 4.5 MWh, in the best-case scenario (i.e. with $x = 2/3$ and the profiles $P_S(t)$ and $Q_S(t)$ optimized), storage enables a gain on line losses of approximately 0.5 MWh. On the other hand, it dissipates approximately 3 MWh at the same time, given its efficiency of 75%. These results agree with those given in [NOU 08], which extends this line of reasoning to a grid including transmission and distribution grids. Using a simplified model of the network, with parameters from the utility company AEP, the authors show that the total reduction of technical losses of the electrical system never compensates for more than 50% of the internal losses of storage systems.

That being said, and although the service is not intrinsically viable, it remains true that certain applications such as peak shaving go along with a reduction of line losses. This may eventually be taken into account in the economic assessment of the storage system, provided that the facility is owned by the grid operator, or that a mechanism that is quite difficult to imagine at present will be developed in order to compensate the operator of the storage system financially for the costs it makes it possible to avoid.

3.3.4.4.3. Power quality

Finally, via the possibilities that certain grid connected power electronic converters could possibly offer, decentralized storage technologies connected by these converters could possibly be used to clean up disturbances of the electrical wave, or "active filtering". The principles of at least two services, which we do not envision except for possible very specific niches, can be defined as:

– *improvement of the quality of electricity supplied to customers*, that is, the use of storage by a distribution system operator to fulfill his commitments with regard to the levels of pollution in the wave voltage he provides to the users of the grid (fast fluctuations, imbalances, voltage dips, etc). In practice, it is usually the customers who pay, more than anyone else, if they like to have access to a level of quality higher than the contractual or regulatory thresholds with which the grid operators provide them; this service will be described later in section 3.3.7.4;

– *improvement of the power quality of electricity extracted from the transmission grid*, that is, the use of storage by a distribution system operator to fulfill his commitments with regard to the levels of pollution in the currents he draws off from the transmission grid. The general principle is illustrated in Figure 3.18 for the example of active compensation of voltage harmonics. This application may, for example, result from the combination of several disruptive sources which, while individually complying with their commitments to the distribution system operators, lead distribution system operators to exceed thresholds on the upstream HV grid.

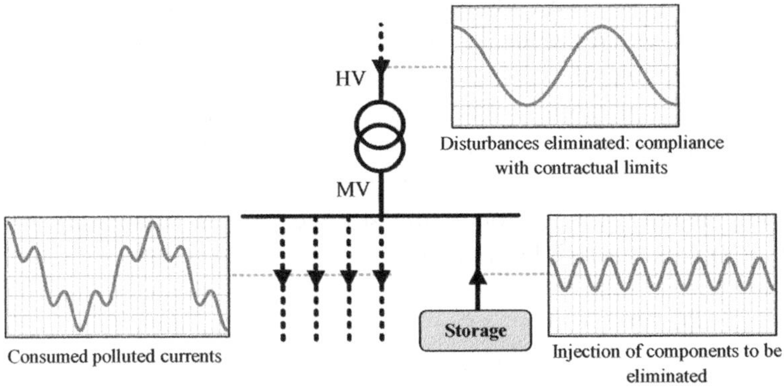

Figure 3.18. *Principle of active filtering of currents drawn off of the transmission grid [DEL 10]*

For these two services that, again, are probably destined to remain theoretical, the cost of conventional solutions could serve as a reference for a case-by-case determination of value.

3.3.5. *Services for a centralized generation owner*

A centralized producer is able to rely on a high-power storage device, for example, pumped hydro, or one based on batteries in addition to its power plants in order to optimize the operation of its production facilities. For this use, parts of the usual electricity generation activities are transferred to storage in order to execute four types of services:

– energy transfer;

– reduction of CO_2 emissions;

– reduction of maintenance;

– provision of ancillary services.

3.3.5.1. *Energy transfer*

Since the efficiency of a thermal or hydraulic power plant depends greatly on the power generated, storage can be used to maximize generation at an optimal economic level of operation. Moreover, energy transfer may make it possible to avoid the use of generation plants with high operating costs. For this service, storage must be sized in order to be able to support daily cycles involving recharge during off-peak hours and discharge during peak consumption hours, with a duration of between 1 and 5 h. High-power and high-energy storage devices such as pumped hydro or compressed air plants are good candidates for this.

3.3.5.2. *Reduction of CO_2 emissions*

Because of the power generation technologies used, the carbon dioxide emissions from a production unit tend to increase during power peaks and transients. Again, for these operating situations, electricity generation can be carried out in part by storage in order to maintain low emissions. The economic interest must be analyzed in terms of generation costs at peak times, which will, in the future, include an ever-increasing CO_2 cost.

3.3.5.3. *Reduction of maintenance*

In addition, storage can support existing generation facilities by avoiding dynamic stresses on old generators. Taking over the most dynamic components of power requests, storage systems can reduce the aging of components, reduce maintenance costs and enable optimal exploitation while complying with operating times and maximum power increases. The requested dynamic characteristics would have a direct influence on the storage technology to be used.

3.3.5.4. *Provision of reserve and ancillary services*

Technically, ancillary services provided to the transmission system operator can be transferred to a storage unit belonging to the producer. Figure 3.19 shows the principle of this type of hybridization on an example involving primary frequency control. Sizing in power, energy and response time depends directly on the provided services. The value of this storage application corresponds to the economic gain in terms of electricity production, which can be exploited with less power limitations. For example, in 2012, in a mining region in Chile, AES introduced a 50 MW Li-ion battery storage system to supply power reserve in place of a 544 MW thermal power plant, which is then able to operate under higher power [AES 14].

3.3.6. *Services for a renewable decentralized producer*

3.3.6.1. *Context and motivations*

The development of decentralized generation causes new scientific and technical problems [ACK 05, GAU 05, ROB 06]. These problems come from new types of sources, distributed geographically and developing rapidly. The commitments made by the European Union to reduce CO_2 by 20% in 2020 compared to 1990 have favored the development of renewable energies for the generation of electrical energy.

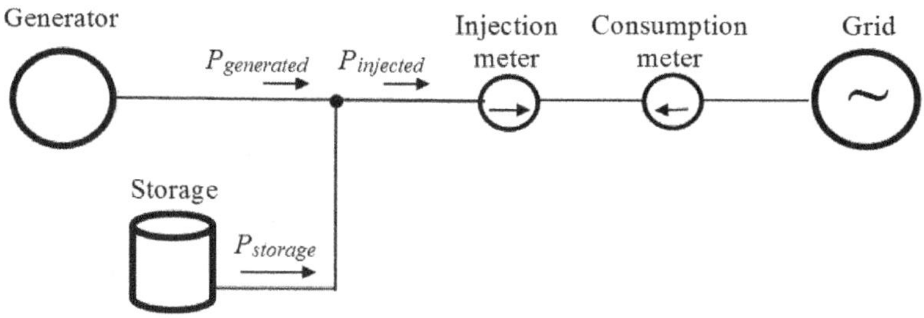

a) For $f < 50$ Hz, injection of a surplus of power ($P_{injected} > P_{generated}$)

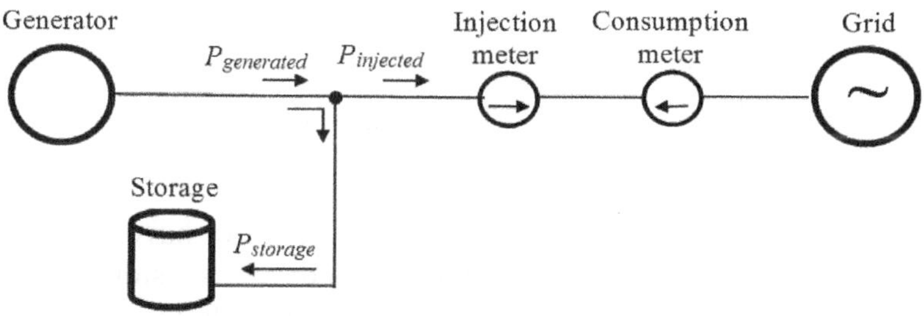

b) For $f > 50$ Hz, reduction of power injected ($P_{injected} < P_{generated}$)

Figure 3.19. *Principle of the takeover of all or part of the frequency control reserve of a generation unit by storage*

Among the known problems, we can cite, for example, the generation of electricity from as-available renewable energies during periods of low consumption. This may lead to a brief overproduction of electricity and even to negative prices reflecting the cost of evacuating this excess energy.

In the example shown in Figure 3.20, the producer has charged its storage system with the surplus energy during the night (in relation to a given maximum evacuable power P_{max}) to inject it during high-consumption hours in addition to its available renewable generation. For a producer exploiting an as-available energy source, the use of storage can make it possible to modulate the electrical power injected into the grid in order to satisfy technical constraints and/or increase the economic value of its product.

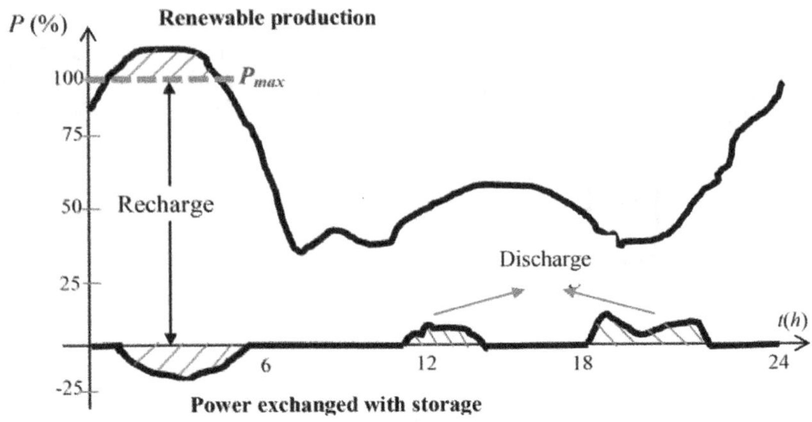

Figure 3.20. *Modulation of the generation profile by energy storage*

In the context of power flow management, we have seen that the transmission system operator may be led to use storage, via the balancing mechanism, as a means of contributing to the resolution of a congestion event on a grid equipment due to the demand to be powered (section 3.3.3.2). In this case, the storage system can eliminate or delay the need to construct new lines and thus reduce the reinforcement time necessary for increased loads. Similarly, for a decentralized producer, the installation of a storage system makes it possible to remove the need to construct new lines and thus reduce the time delay necessary for the connection of additional generators.

From an economic point of view, the electricity market presents risks for renewable generation due to penalties in the event of a negative difference between the amount of power announced/sold and the amount actually generated. The difficulty of forecasting generation may result in a significant shortfall, which would then be added to the shortfall linked to limited correlation between the available generation and the load. Storage can provide services that would ease the integration of decentralized renewable generation into the market.

These services are briefly introduced in the sections below.

3.3.6.2. *Injection deferment*

Some renewable generators, such as wind and photovoltaics, have a fluctuating primary resource that is difficult to predict. In fact, it can occur that the renewable primary source is abundant when the charge is low or, conversely, that the primary source is absent at the peak load time. From an economic point of view, if the energy supply is not available during periods of high demand in a context without feed-in tariffs, there is a risk of low sale prices and thus a shortfall for the producer.

The use of storage to defer an injection can make it possible to maximize the economic value of the energy produced while also being of interest for electrical systems, as a substitution for costly and polluting peak-time resources. Injection deferment consists of charging an energy block during the least economically interesting times of the day with a view to a use later when its price is higher. The value of this operation lies in the difference in gain between a "real-time" injection of the energy and a partial or total deferment. These applications are available according to whether renewable energy is included in a market or is paid by regulatory incentive rates.

Figure 3.21 illustrates the injection deferment service in the context of wind generation that is part of an energy market. During the lowest consumption times between 2:00 and 6:00 am, the storage system is recharged with surplus wind energy, which is then discharged during the peak morning hours between 10:00 am and 2:00 pm, when the market price is higher.

As part of regulated rates, storage can also be used in the development of self-consumption [GER 08]. The difference in comparison to a plant with total sales is that all or part of the renewable energy-based generation is consumed by the producer–consumer without being injected into the grid (Figure 3.22). The principle of self-consumption is to power an electrical facility by means of its own generation at the time it is produced by the renewable electric generator. This is typically the case with photovoltaic panels, for which sunshine varies throughout the day and generation thus undergoes peaks and dips. To carry out self-consumption, it is necessary to try to correlate periods of consumption with periods of generation. This can be done by programmable loads or devices whose start-up is triggered by detecting the availability of renewable generation. However, this type of electrical load remains limited, and it appears more flexible to store electrical energy locally when it is available, in order to consume it when the need arises.

a)

b)

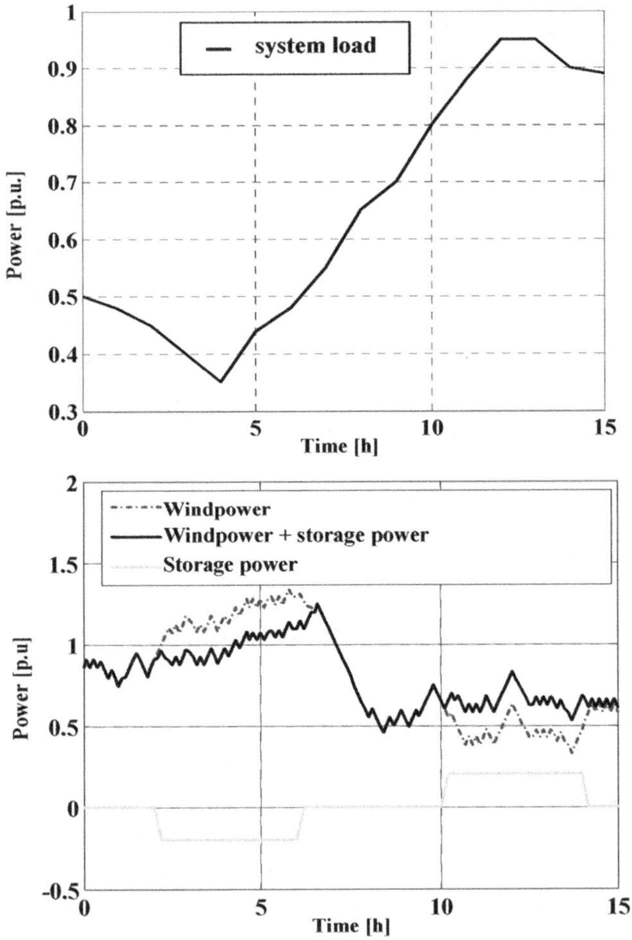

Figure 3.21. *Illustration of injection deferral for windpower production*

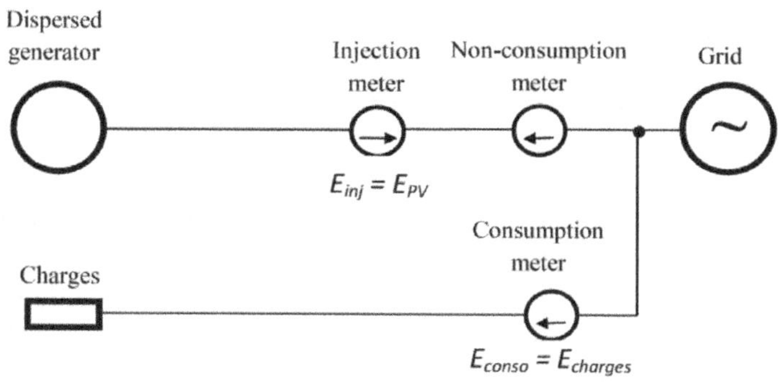

a) Metering for the sale of total production

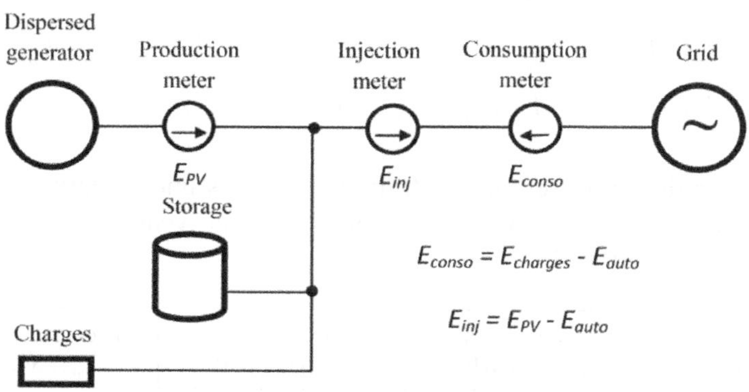

b) Metering for self-consumption, which can be reinforced using a stor:

Figure 3.22. *Comparison of metering approaches between total sale and self-consumption of renewable generation*

For a given producer–consumer, the interest of self-consumption appears only if the sale rate of production T_{inj} (c€/kWh) is lower than the rate of consumption T_{conso} possibly increased by a self-consumption premium T_{auto}. The introduction of a storage system may be justified in the event that a significant difference encourages maximal self-consumption. In the simple example shown in Figure 3.22, the dispersed generator produces an annual quantity of energy E_{PV} and loads consume an annual quantity of energy $E_{charges}$. In case A, sale of total production, the net annual bill of the producer–consumer can thus be written as follows:

$$FactA = E_{conso}T_{conso} - E_{inj}T_{inj}$$
$$FactA = E_{charges}T_{conso} - E_{PV}T_{inj} \quad [3.12]$$

In case B, the producer–consumer self-consumes a volume E_{auto} of its total production E_{PV}, which remains unchanged. Disregarding energy losses in the storage (if any), the energy it extracts from the grid E_{conso} thus becomes equal to $E_{charges} - E_{auto}$; at the same time, the energy it injects into the grid E_{inj} has a value of $E_{PV} - E_{auto}$. Thus, its net annual bill becomes:

$$FactB = E_{conso}T_{conso} - E_{inj}T_{inj} - E_{auto}T_{auto}$$
$$FactB = \left(E_{charges}T_{conso}\right)T_{conso} - \left(E_{PV}T_{inj}\right)T_{inj} - E_{auto}T_{auto} \quad [3.13]$$
$$FactB = FactA - E_{auto}(T_{conso} + T_{auto} - T_{inj})$$

For the purposes of illustration, the rates under German law for 2008 are shown in Table 3.6 according to the year of installation of the system [RIF 09].

Year of installation	Consumption rate T_{conso} (c€/kWh)	Injection rate T_{inj} (c€/kWh)	Self-consumption rate T_{auto} (c€/kWh)
2014	22.04	27.13	15.78
2015	22.70	24.69	14.36
2016	23.38	22.47	13.07
2017	24.08	20.45	11.89

Table 3.6. *Rates according to 2008 German law [RIF 09]*

In this example, the price of electricity is estimated on the basis of a 3% increase per year. Self-consumption is favored because, in each case, we have:

$$T_{conso} + T_{auto} > T_{inj}, \text{ in other words FactB} < \text{FactA} \quad [3.14]$$

3.3.6.3. *Guarantee of a production profile for buyers and the system operator*

If renewable producers are part of the electricity market, they must define an hourly production plan one day ahead based on their predicted generation, and keep their commitment subject to penalties in the event of a discrepancy between actual generation and the generation that was announced. The uncertainties of predicting the behavior of primary sources (wind speed, sunshine, etc.) complete this process and result in errors in the estimation of the electrical generation that can be achieved upstream of real time. According to the literature, forecasts are correct overall in terms of magnitude and volumes, but become uncertain about the time when this level of generation will be obtained [HOL 04]. In Denmark, error rates for generation made 24 h ahead regarding wind production have sometimes reached 50% due to the difficulty of defining peak generation times [ACK 05].

In this context, storage systems could smooth generation variations while guaranteeing compliance with levels announced in advance [LU 10a, LU 10b]. The sizing of storage for keeping to a pre-established program varies, in particular with regard to the characteristics of the product of the source to be guaranteed, and to the reliability with which its evolution can be predicted [KOE 08]. For a given renewable energy source, this involves great dependence of the results on geographic criteria and on the time considered for the prior declaration of the generation plan. A practical example of this type of service is the Rokkasho–Futamata project in northern Japan, which since 2008 has used 34 MW/204 MWh of NAS batteries to match the production of a 51 MW windpower farm to a stair step profile.

3.3.6.4. *Contribution to ancillary services*

In France as of 2014, electrical generation plants using renewable energies such as wind and photovoltaics were not required to participate in frequency control. However, in other electrical systems such as in Ireland, a constructive capacity to participate in primary frequency control is mandatory. In such cases, as shown in Figure 3.6, wind farms must keep a permanent power reserve, which causes a shortfall since any generation that is not used when it is available is lost. As part of an obligation to contribute to frequency control, wind producers may choose to rely on a storage system to supply the required power reserve rather than on their turbines. In this way, they can continue to produce the maximum available power and can guarantee the availability of the primary reserve regardless of wind conditions.

As shown by [DEL 10], because storage makes it possible to avoid a loss of the energy produced, the value making of this service is directly related to the value of the energy produced by wind farms. In the case of France, where the regulated price of windpower is on the order of €80/MWh, Delille [DEL 10] calculates, for the purpose of illustration, that the value making may be equal to €160 k/year per MW of storage installed. In these conditions, the use of a storage system to carry out primary control instead of renewable producers seems economically viable. However, it is of less interest as the price per renewable kWh is dropping.

3.3.6.5. *Value making of curtailed energy*

The periodic curtailment of real power is requested by the grid operator when it is not possible to inject electrical energy into the grid at the time it is available, for a given reason such as distribution or transmission congestion, limited penetration rate for dynamic reasons, voltage constraints, etc.

To avoid a loss of generation, the value making of curtailed energy consists of the storage/transfer of available energy at the time of the curtailment request (Figure 3.23). With a storage system, the producer can accumulate this energy in order to sell it later, minus a share of losses (see [ABO 05, BAR 04] and [EPR 05b]). The value making of this service comes from the exploitation of an energy product that would have been lost without a storage system. The price of energy at the time of reinjection remains up for debate, specific purchase price, market price, etc. In any case, costing depends on the frequency of curtailment, which suggests a case-by-case study. The sizing of the storage system also remains to be evaluated and should be relatively massive, in particular if there is a question of absorbing the nominal power of the renewable source during curtailment. A connection of the storage system within the power plant concerned and downstream of the metering point seems the most appropriate for rendering this service. Other configurations may be imagined, however, to maximize the possibilities of mutualization of services, with the only limit even for theoretical propositions being that they must be able to act on the limitation that is at the root of the need for curtailment.

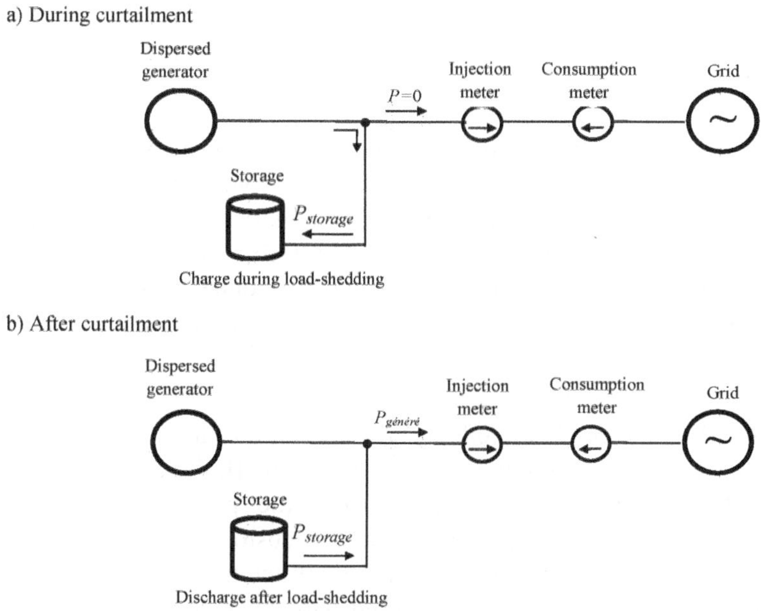

Figure 3.23. *Illustration of power flows during and after curtailment*

Besides the sale of the energy produced during disconnections requested by the grid on system operator, there are other possible sources of value for generation curtailment. In particular, the producer's agreement to reduce its real power from time to time may make it possible to postpone or avoid reinforcements of the electric grid at the time of the connection request, which would have the advantages of (1) reducing the costs to the producer at the time the plant is set up (and potentially the costs to the grid operator as well) and (2) reducing connection time, in particular if certain new works come up against environmental or administrative constraints.

For this service, in terms of energy management of the storage system, it is theoretically desirable for the energy to be discharged before any curtailment event in order to minimize the loss of renewable energy.

3.3.7. *Services for consumers*

The services discussed briefly below concern mainly industrial or commercial customers. To a lesser extent, some applications may be of interest now or in the future for individual consumers, even though the additional complexity, size and possible risks involved might potentially dissuade them. Recall that the contribution of storage to domestic self-consumption was discussed in section 3.3.6.2.

3.3.7.1. *Peak-load shaving*

Of the services intended for consumers, the literature places great emphasis on peak shaving (see in particular [EYE 04, EYE 10] and [NOR 07]), whose value arises from the principle of electricity pricing. The price actually billed to the customer is the sum of one part proportional to the power contracted for, and one part proportional to the energy consumed. However, the power contracted for is a maximum value liable to be drawn off, which is often reached only marginally in reality. From the consumers' point of view, this goes back to booking a service and then underutilizing it or, in other words, to pay too much for what is actually needed.

Peak shaving consists of shaping a customer's consumption profile in order to reduce the power contracted by that customer, and thus the amount billed; however, this is at the cost of an increased drawing-off of energy (losses from storage). To do this, as shown in Figure 3.24, recharging is done when demand is low, and discharge is synchronized with peak power times.

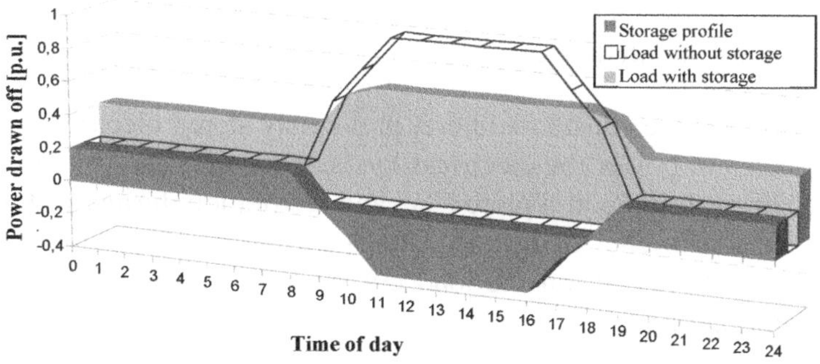

Figure 3.24. *Peak shaving using a customer storage system [DEL 10]*

The interest of this type of operation largely depends on the form of the customer's load profile and various aspects such as the cost of the power contracted for, and the details of billing for possible overspending. The value of peak shaving is discussed by [EYE 04] and [NOR 07] in an American context. According to references [MAR 98] and [OUD 06], the most favorable cases correspond to short peak consumption times that can be predicted in advance, in particular in order to limit the storage capacity to be installed. Each of these two sources presents a case study showing that peak shaving could be profitable under some conditions in an industrial context.

Mutualizations of services are possible to enhance the value of a storage put in place to perform peak shaving, such as contributions to quality/continuity or reactive power compensation. Moreover, this service is liable to cause a consumption deferment (see section 3.3.7.2) if the peak occurs during full-price hours.

3.3.7.2. *Deferment of consumption blocks*

This service is aimed at reducing the "energy" part of the electricity bill of a customer taking advantage of hourly differences in prices. Historically, the service was developed to shift the demand for electricity by storing energy during low-demand periods. It is implanted via the control of electric water heaters which store the electricity consumed in the form of heat for future use. Extensions based on controlling the recharge of electric vehicles have been suggested by the literature.

For the service considered here, previously stored electrical energy is reinjected to power the electrical loads. Accumulation is carried out during off-peak times at a price C_{HC}, and restoration during peak times at a price C_{HP} higher than C_{HC}. The operating profile of the storage system and its impact on the load curve of the consumption site are shown in Figure 3.25.

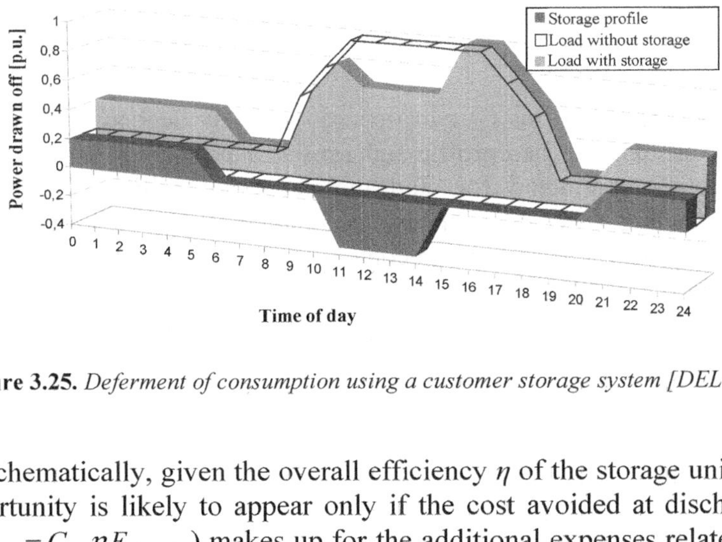

Figure 3.25. *Deferment of consumption using a customer storage system [DEL 10]*

Schematically, given the overall efficiency η of the storage unit, an opportunity is likely to appear only if the cost avoided at discharge ($\Delta C_{HP} = C_{HP}\eta E_{chargée}$) makes up for the additional expenses related to the purchase of stored energy ($\Delta C_{HC} = C_{HC} E_{chargée}$). In mathematical terms, we write $\Delta C_{HP} > \Delta C_{HC}$, meaning that the ratio of peak-time prices to off-peak-time prices must verify $C_{HP} / C_{HC} > 1/\eta$ for a consumption deferment to start being pertinent.

The value of this service depends mainly on the price scheme applied by the supplier. With a significant peak/off-peak difference ($0.37 to $0.11/kWh), the calculations made by [EYE 10] lead to a revenue of $167/year per kW of storage installed, assuming 6 h of discharge time and an 0.8 efficiency η. In the case of France, with the peak/off-peak ratio being much lower than the 3.4 factor considered by this reference, the potential gain is relatively modest; simple calculations can show that it might at best make up for the cost of losses in the storage unit. That being said, the possible benefits could increase in the future with the appearance of contracts reflecting peak-time effects.

As deliberately shown in Figure 3.25, consumption deferment does not necessarily involve peak shaving, though the two services can easily be mutualized in a scenario where a customer's peak consumption time coincides with peak hours.

3.3.7.3. *Participation in demand response programs*

According to a number of stakeholders, consumers might be called in the future to participate increasingly actively in the operation of the power system, particularly by adjusting their consumption to assist with the maintenance of frequency, voltage, the optimization of the use of generator plants or the resolution of congestion events. These are advanced forms of load control that would be in addition to the electric water heaters with storage already in service in France for many years.

A customer equipped with a storage unit would contract for a periodic load shedding service with a grid operator or a supplier. The interest for the customer would lie in being able to make its contribution with a minimum impact on usual operations, with its consumption remaining guaranteed to a certain extent during times of increased stress. Pricing this use of a storage system is difficult, since it depends partly on the amount of benefits the customer may receive from its participation in demand response schemes, and partly on the value it assigns to the maintenance of its normal consumption modes.

In any case, this service involves the indirect exploitation of a number of the storage values described earlier, such as the investment deferral for distribution system operators, or frequency and voltage controls for which the transmission system operator is responsible. From the customer's perspective, mutualizations are undoubtedly possible with "peak shaving" or "consumption block deferment"-type applications, with these needs likely to appear during times of heavy charge in the electrical system.

The necessary power is a fraction of the power contracted for by the consumers concerned, from several kilowatts to several megawatts, for discharge times varying from a few minutes to a few hours depending on the need motivating the load reduction request.

3.3.7.4. *Specific requirement for power quality*

Storage systems can be used to filter short-term disturbances coming from the grid in order to protect equipment that requires a specific quality level. This may, for example, involve eliminating voltage drops to protect a sensitive appliance.

The value making comes from the losses of productivity avoided by the enhancement of quality, which means that it can only be priced on a case-by-case basis. The storage unit is connected directly, as close as possible to the loads to be protected, or possibly electrically near if the operator of the storage system discovers that there are additional possibilities in terms of the mutualization of services, and satisfactory contractual terms can be defined.

3.3.7.5. *Continuity of power supply*

A storage system can be substituted opportunely for the grid in the event of periodic cuts from the power supply. This service has been known and exploited commercially for many years to reduce consequences and power outages in industrial applications. In some cases, damage is caused by the occurrence of the power outage itself, whatever its duration (e.g. data losses from servers). The solution consists of having access to highly dynamic resources able to take over the principal supply of power for a period of time. The emergent safety supply can be extended in time or simply interrupted after having enabled the controlled shutdown of certain equipment and the completion of backups. In other cases, it is the duration of the power outage that is the main problem (for instance, in the case of a cold chain); what matters in this case is to have access to an emergency supply system that can be started without necessarily a limited response time, but with a certain guarantee of independence.

The value of this service is linked to the economic losses avoided via the strengthening of continuity and depends on the specific characteristics of each situation [EYE 04, EYE 10]. Economic assessment can be difficult for domestic customers (for details, see [DOU 02]).

There are numerous commercial products on the market for uninterrupted power supplies, from fossil fuel generators to storage systems using various technologies. For this service, the storage unit is connected directly to the consumer's appliances, or could possibly be shared among several neighboring customers, for example, on the scale of an area of activity, as is the case of support by intentional islanding, which we discuss in more depth in section 3.3.4.3.

3.3.7.6. *Limitation of disturbances caused on MV or LV grid upstream*

Customers agree, as part of the contract for access to the distribution grid, to limit the disturbances generated by their facilites. Without going into further detail, we will simply mention that current clauses in France concern fast voltage fluctuations (fits and starts, flicker), imbalances and harmonics. If a customer drawing-off power at MV does not respect the limits imposed on it, it will be billed for the costs involved in corrective decisions taken by the distribution system operator. As discussed in detail in section 3.3.4.4.3, decentralized storage can contribute active filtering solutions and thus prevent the propagation of disturbances on the grid.

The value making of storage for this service corresponds either to the cost of the decisions taken by distribution grid operators to resolve the problem of lack of quality, or to the cost of the conventional solution that would have to be installed by the customer. On that note, there are also a large number of commercial products dedicated to these applications, some of which are well managed and competitively priced (e.g. passive filters), and undoubtedly leave little room for possible competitors that are more innovative but also more expensive. Typically, with regard to harmonics, active filters are considered only in extreme cases where all other alternatives have been exhausted.

With regard to specifications, the power levels to consider are between some 100 kVA and approximately 10 MVA, with a few seconds of discharge at the most, and a fast dynamic.

3.3.7.7. *Compensation of reactive power*

Users powered at MV or LV (more than 36 kVA) are billed for the reactive energy consumed beyond a tangent *phi* of 0.4 during certain periods of the year. Under the terms of TURPE4, the amount to be paid is, for example, approximately €1.8c/kvarh for this term of reactive power.

Because of the possibilities offered by its rotating or static interface with the grid, a decentralized storage system could compensate locally for the reactive power consumed by loads. A stakeholder equipped with a storage system could take advantage of this additional service to help make its storage plant profitable. The amount of economic gain can be established on the basis of the reactive power cost noted above, or on that of the capacitor banks that would enable the customer to fulfill its commitments to the operator of the grid to which it is connected.

3.3.8. *Benefits from market activities*

3.3.8.1. *Purchase and sale of energy blocks*

One application that has been widely discussed in the literature consists of adapting the energy management of a storage system according to market electricity prices; accumulation is carried out when prices are at their lowest, to be recovered when they are at their highest. This service is usually referred to rather erroneously as "arbitrage", a term used in finance for purchase/sale transactions that are simultaneous rather than successive.

The economic aspects of "arbitrage" are discussed in sources [EYE 04, EYE 10] as well as [IAN 05], based on data taken from American markets. The operation is of interest only if the difference between the resale price and the purchase price of energy (including the cost of delivery by the grid) is great enough to cover the investment and operation of the system.

To estimate the value of this type of service in the French case, the maximum gain on a series of Powernext 2007–2008 spot prices is estimated in [DEL 10]. To do this, the optimal daily operating profile for a 1 MW storage system has been calculated, with conversion into equations done using the following approach:

– Each day of the year is treated like an independent problem whose objective is to maximize the profit resulting from market transactions by adapting the variables constituted by the 24 hourly reference levels of active power of the storage system.

– Constraints have to do with the state of charge, which must stay at all times within the usable capacity range of the system and be zero at the 24th hour, with all the energy loaded being released before the end of each day.

Figure 3.26 shows the results obtained, taking into account, for example, an indicative delivery (grid) price of €10/MWh and a high overall storage system efficiency of 80%. These calculations give us an annual value of the storage system ranging from €25k/MW (1 h of discharge) to €75k/MW (7 h of discharge), for an present revenue at 8% over 10 years of €170k–500k/MW, which clearly seems inadequate given the investment costs of the storage system.

Moreover, this type of approach carried out retrospectively on recorded prices tends to maximize the value resulting from operations carried out compared to what can be recovered in practice with a positioning on the market at D-1; the amounts given should be considered as upper bounds.

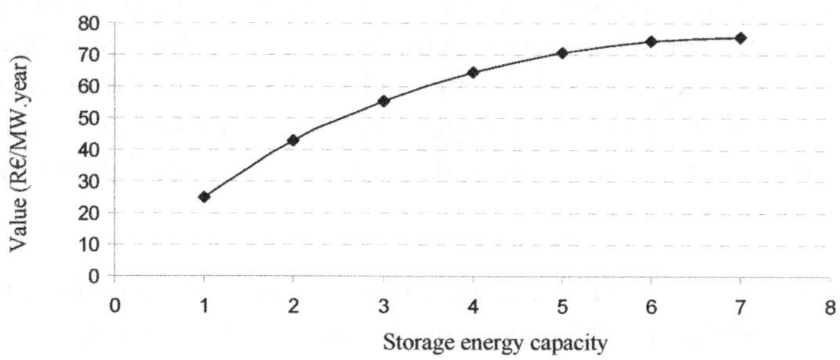

Figure 3.26. *Value creation of storage for a year of energy arbitrage [DEL 10]*

These estimates show that it is difficult to make a profit in current conditions by basing the value of a storage system on this service alone. That being said, daily price evolutions may offer an additional source of value for applications using an adapted operating profile, that is, a storage during off-peak times for deferment to peak times. With regard to sizing, access conditions require power levels greater than 1 MW for the wholesale market; this value can be obtained via aggregation (see [HAR 05] in particular).

3.3.8.2. *Ancillary and system services market*

In the past, services vital to the power system security and reliability, such as frequency control, were provided exclusively by the production plants connected to the transmission grid. A current trend is to offer the possibility of providing these vital services by other equipment so that the transmission grid operator can contract for maximum flexibility and with more categories of grid users.

In France, participation in frequency control was opened up in 2014 to consumers and aggregators for experimentations at first. Only consumption sites with a contract for access to the distribution grid (Contrat d'Accès au Réseau de Distribution (CARD)) and those that are hosted, or indirectly connected to the public transmission grid, are concerned. The stakeholders controlling these sites can sell their primary or secondary reserves at an unrestricted price to producers, who can use it to fulfill their obligations to RTE. This activity will also be open to traders beginning on 1 July 2015. This promises the emergence of a frequency control market on which primary and secondary frequency reserves will be able to be exchanged. In Belgium, frequency control via adjustable charges has been in service since 2013. In this context, storage, like charge aggregation activities, will be able to be used to provide services commercialized on new "ancillary service markets".

With regard to tertiary reserves, new possibilities are also opening up. For example, in Belgium, the transport grid operator Elia has access to a tertiary reserve supplied by the interruption, if necessary, of large consumers, which was augmented in 2014 by the load-shedding of small consumers connected mainly to the distribution grid. In this context, for a consumer having subscribed to this type of service contract, storage can be used to power part of its domestic loads during periods when this service is requested. This is the application described in section 3.3.7.3, sized according to loads to be powered and, for the current Belgian case, for a duration of 8 h for 4–12 requests per year.

3.4. Example of the contribution of storage to the treatment of congestion events

3.4.1. *Indicator of state of charge of grid*

The state of charge of the grid represents the distribution of the power flow, which depends on the location of the production and consumption and the means of compensation for reactive power and for impedances of transport infrastructures. Congestion appears when the physical limit of hardwares is exceeded, defined by IMAP (see section 3.2.4.4). However, in addition to IMAP, safe operating limits are considered, as the operation of electrical lines at their maximum capacity can pose risks, in particular the risk of cascading overloads. As shown in Figure 3.27, the flow of power between two zones linked by three identical lines is distributed uniformly over all of these lines. In the event that a line is triggered, the power flowing on the two remaining lines increases by 50% on each line.

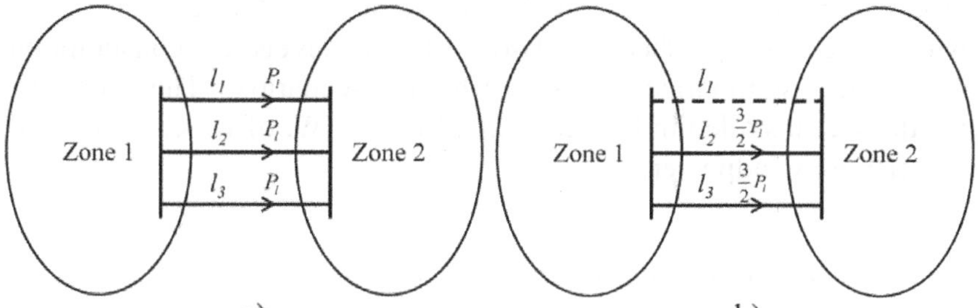

Figure 3.27. *Distribution of power flows in the case of three identical lines. a) All three lines are present and b) after line l_1 is triggered*

If all three lines were at their maximum capacity at the time of triggering, the two lines remaining will be overloaded after the incident. To guard against this phenomenon, a safety margin is necessary. The N-1 rule is frequently used to quantify this margin; it makes it obligatory for the grid to remain operable in the event of the loss of any element of the grid. The safety margin is such that after the loss of a mechanism, no physical limit can be exceeded on the remaining mechanisms. However, when multiple electrical systems are interconnected, this margin is difficult to define, as it depends on the latest injections and drawing-off actions of all of the electrical systems. To identify congestion zones at N and N-1, an indicator of the state of charge of the grid (ITC_l) has been defined [VER 11b]:

$$ITC_l = \frac{S_l}{Smax_l} \qquad [3.15]$$

where $S_{max,l}$ and S_l are the apparent maximal and real power levels of a mechanism l. If ITC_l is greater than 1, a congestion event is indicated.

3.4.2. *Evolution scenario for electric grid*

To illustrate this example of the contribution of storage, three scenarios for the evolution of the French electric grid in 2020 in terms of peak consumption, classic power production and windpower production are compared with two states of the grid in 2002, which are distinguished by two different levels of consumption. Table 3.7 summarizes these different scenarios. The goal for windpower installed in France in 2020 is 25 GW, of which 5 GW will be offshore windpower.

3.4.3. *Treatment of congestion events in Brittany*

The application of the state of grid charge indicator to the French grid makes it possible to identify areas where congestion events are most at risk of appearing, specifically in the regions of Brittany and Provence-Alpes-Cote d'Azur.

Operating point	Year	Period	French consumption	Nuclear, thermal and nuclear facilities in operation	Effective windpower production
A	2002	8/21 at 11:00 am	48.9 GW	82.5 GW	0 GW
B	2002	1/16 at 11:00 am	69.5 GW	82.5 GW	0 GW
C	2020	Third Thursday in August between 3:00 and 5:00 pm	35.3 GW	93.7 GW	17 GW
D	2020	Peak load in winter between 7:00 and 8:00 pm	90.8 GW	93.7 GW	17 GW
E	2020	Peak load in winter between 7:00 and 8:00 pm	90.8 GW	93.7 GW	0 GW

Table 3.7. *Summary of operating points*

Thus, for example, congestion events appear in scenarios B, D and E on the Brittany grid, which is shown in Figure 3.28. Congestion events appear on lines 1, 2 and 3 following the loss of the 400 kV line linking nodes 94 and 113. The values of the indicator are given in Table 3.8.

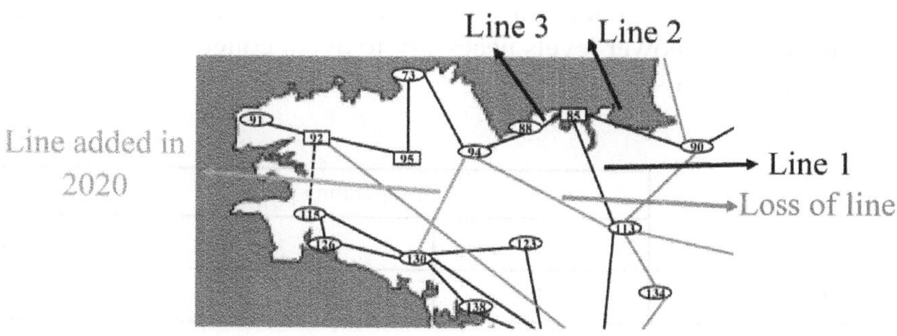

Figure 3.28. *Studied Brittany grid. For a color version of the figure, see www.iste.co.uk/robyns/powergrids.zip*

Overloads are high due to high local consumption and the lack of local production, which requires a power flow to feed the load. Comparing scenarios B and D, there will be fewer congestion events in 2020, since the addition of a line between nodes 94 and 130, which is in progress, will relieve the grid in northeastern Brittany. Moreover, the presence of additional windpower and the 500 MW offshore wind farm off Saint-Brieuc, connected at node 94, will contribute to reduced congestion.

Indicator according to lines	Operating point		
	B	D	E
ITC1 (%)	115	108	140
ITC2 (%)	125	115	157
ITC3 (%)	130	128	182

Table 3.8. *Indicator of the state of charge of the grid*

Case E, in which wind production is considered to be zero, is the most critical scenario with an overload of 82% on line 3. The congestion events appearing can be avoided by the connection of a storage system at node 88. With regard to the sizing of this mechanism, the power levels necessary to avoid congestion events are shown in Table 3.9.

	Operating point		
	B	D	E
Power (MW)	250	178	386

Table 3.9. *Power required to manage the congestion events with the aid of a storage system connected at node 88*

3.5. Example of contribution of storage to dynamic support of frequency control in an island grid

3.5.1. *Context and potential interest of this service*

Because of their excellent dynamics, the contribution of new storage technologies to frequency control has been a subject of interest since the 1980s, with the installation of a 17 MW/14.4 MWh lead–acid battery to increase the reliability of the electrical power supply in the isolated system of West Berlin [KUN 86]. More recently, new technologies such as carbon fiber flywheels and lithium ion batteries have been used for similar applications in islands including Hawaii, and even within interconnected grids, as in the United States.

In this section, we will introduce a new potential service for storage devices with short response times, specifically dynamic support for frequency control, whose objective is to reduce the use of under-frequency load shedding in island systems. As we will see by applying this service to the case of the Guadeloupe system, an application like this makes it possible to take advantage of the dynamic performances of certain storage systems and to increase their value from the point of view of the electrical system, which may help encourage their development in the long run.

3.5.2. *What is under-frequency load shedding?*

When the primary reserve is not enough in the face of a sudden imbalance between supply and load demand, or if its release dynamic is too slow, customer shedding is the only way to slow down the loss of kinetic energy in conventional generator plants and thus to stabilize frequency before generation plants disconnect preventively from the grid. The automation of under-frequency load shedding is mainly intended to avoid widespread incidents when possible [RTE 04].

In practice, this function is provided within primary substations. Consumption is divided into different subsets, called echelons, in order to disconnect only the volume of load necessary to re-establish equilibrium. For example, in continental metropolitan France, MV feeders are divided into five echelons, each representing approximately 20% of the total power. In the event that a frequency threshold is exceeded, the automated system instantly disconnects consumers according to an order of priority of power supply, beginning with echelon 1, which is controlled at 48.5 Hz in Guadeloupe at present. The final echelon, which includes priority users, cannot be shed; it ceases to be powered in the event of a blackout. Following the incident that necessitated the activation of under-frequency load shedding the charges concerned are progressively reconnected by the grid operator via manual control.

3.5.3. *Technical specifications of dynamic support*

A low-power level and the lack of interconnection create islanded electrical-systems-specific characteristics. Of these, we would note the low kinetic energy of weak rotating masses, and the large unitary size of the generation units compared to the total power produced. These two properties, when combined, result in high-frequency variability, particularly in the case of the sudden loss of a generation unit. In addition, the growing insertion of sources using renewable energies tends to aggravate this phenomenon; these new productions take the place of conventional generator plants, contributing little (in the case of wind technologies) or no (in the case of all photovoltaic installations) inertia to the electrical system.

In the PhD thesis [DEL 10], the analysis of a 3-year history of faults on the archipelago of Guadeloupe (2006–2008) shows that up to 50% of the under-frequency shedding can be explained by a primary power reserve dynamic release criterion. In other words, for these incidents, the available power is sufficient, but the response time of conventional electricity production technologies and of their controls is inadequate in the face of the frequency gradients observed in practice. It is in this context that the performances of certain storage technologies may provide an interesting support to the dynamic behavior of islanded grids.

The principle of dynamic frequency control support is to reduce the depth of the dip after an unexpected event while giving conventional resources enough time to increase power levels. To do this, this service relies on an impulse reserve which temporarily completes the primary control action, whose characteristics are as follows:

– its deployment must be as short as possible, in the order of a second at most, to be useful in a context where frequency can drop by more than 1 Hz/s;

– its holding time must cover the attainment of the minimum frequency (a few seconds) and can extend to covering the complete deployment of primary control (approximately 15 s);

– its termination of action must be progressive enough not to create a sudden new imbalance between production and consumption, which can be harmful to the safety of the electrical system.

Figure 3.29 locates this service in relation to currently existing controls in French island grids, specifically, as shown in Figure 3.28(a), automatic primary control that acts initially to stop a frequency drop and the manual actions of the dispatcher, who then takes over to re-establish the situation, for example, by starting up a gas turbine. With dynamic support, which intervenes as shown in Figure 3.28(b) during the activation of the primary reserve, part of the imbalance is compensated for by the impulse reserve instead of taken from the kinetic energy of the generator groups. Consequently, the maximum depth of the frequency drop following an incident is reduced, which can make it possible to avoid the activation of metric-frequency shedding on a criterion of dynamic release of the primary reserve. If the reserve available in the conventional power production system is sufficient, the frequency will stabilize under the effects of primary control alone after the termination of the dynamic support action.

Obviously, if the sizing of the storage system allows it, the injection of power can be maintained beyond the temporary system, in addition to or as a replacement of the generator reserve. This is a contribution to primary control, or even more, if it is possible to continue the supply on a long-term basis; these are other services whose technical and economical aspects have already been the subject of practical studies and projects, as discussed earlier in this chapter.

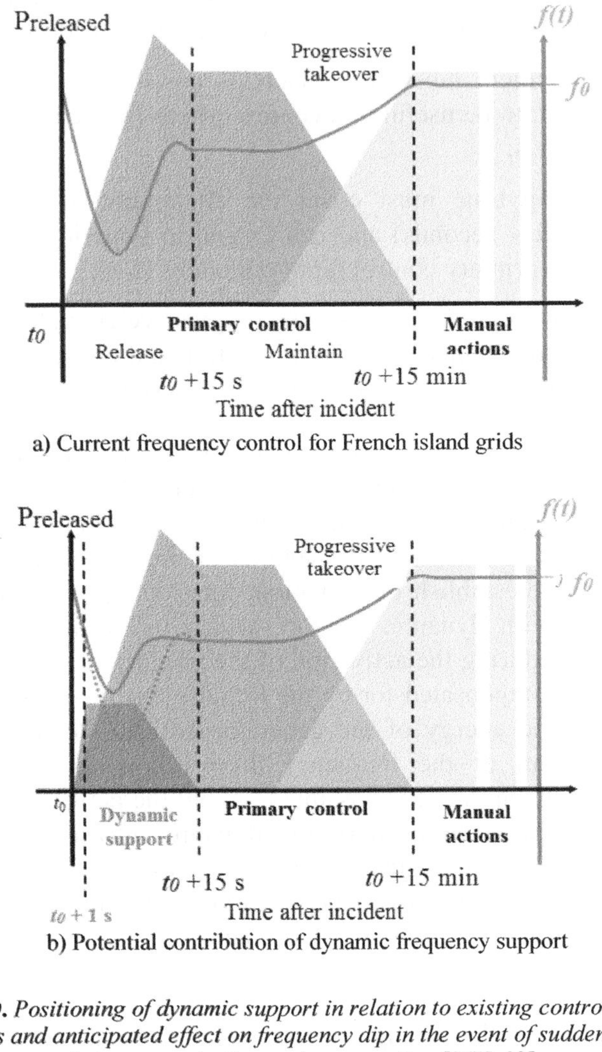

Figure 3.29. *Positioning of dynamic support in relation to existing controls of French island grids and anticipated effect on frequency dip in the event of sudden imbalance between production and consumption [DEL 12]*

3.5.4. *Method used for detailed study of dynamic support*

The technical and economic study of this service contributed by storage is based on a determinist approach, that is, it is based on worst-case scenarios. Three simulation scenarios for the Guadeloupe system simulation have been applied:

– scenario 1 is based on the most critical unforeseen loss of a production group in the 2006–2008 history mentioned above;

– scenario 2 is based on a real configuration observed in 2009 with an instantaneous penetration rate of 12% of variable renewable energies;

– scenario 3, developed on the basis of scenario 2, is a fictional case with an instantaneous penetration rate of 29% of variable renewable energies.

This service can be studied in three stages, which are presented below: (1) theoretical approach, (2) dynamic simulations and finally (3) experimental laboratory implementation.

3.5.5. *Stage 1: theoretical approach*

In this section, an initial sizing of the storage system for dynamic support is attempted using a theoretical approach based on a variational model of the electrical system (standard "small signal" approaches), which is extended to estimate the storage power necessary for the service studied in islanded grids. The hypotheses considered are as follows:

– The grid modeled is limited to a single synchronous zone within which the frequency is supposed to be uniform at all times. "Small signal" simplifications are taken into account in the conversion into equation of the evolution of the rotation speed of the production generator plants.

– The dynamic behavior of all units contributing to primary frequency control (speed controller, actuator and the processes included) is represented in the form of a unique transfer function. This choice is well justified in islanded grids, where the majority of reserve power is often provided by power plants with a single technology, whose characteristics therefore determine the dynamics of the system. The form used to approach the usual models of a higher order in a basic manner is a lead and lag filter.

– All of the nonlinearities of power plants or their primary control are disregarded (deadband, variation of droop and dynamics according to operating point, saturations due to control and equipment, etc.). In particular, the calculations do not contain a power limiter; the available reserve power is therefore systematically sufficient.

3.5.5.1. *Model of a system isolated around a point of equilibrium*

According to classic reasoning under the "small signals" hypothesis, which is described in [KUN 94], for example, the evolution of the frequency f of an electrical system starting from a given initial state can be described by the following equation, and the associated Figure 3.30. In these representations:

– M_{eq}, in seconds, is the characteristic time related to the inertia of the system, or "mechanical time constant";

– D, without measurement, is the charge self-regulation coefficient, which measures its sensitivity to frequency variations;

– ΔP_m, in per units of a given base VA_{base}, is the variation in the mechanical power produced by the generators;

– ΔP_{ch}, in per units of a given base VA_{base}, is the variation in the electrical power required by loads independently of the evolution of f:

$$\Delta f(s) = \frac{1}{M_{eq}s + D}(\Delta P_m(s) - \Delta P_{ch}(s)) \qquad [3.16]$$

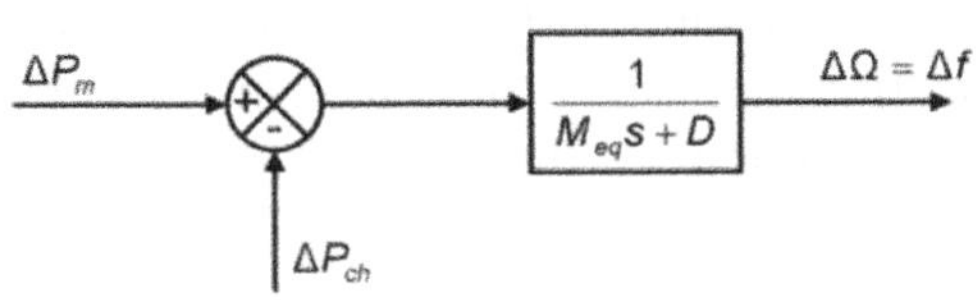

Figure 3.30. *Simplified dynamic model of an electrical system around an equilibrium point [KUN 94]*

Upstream of the schema in Figure 3.30, the power on the equivalent shaft of the n' machines participating in primary control on the n in operation ($n' \leq n$) is generated by a speed control loop acting on an actuator, which manages the power supply of a turbine or an engine driving the mechanical coupling. The calculations presented here are aimed at approaching the dynamic behavior of the system via a literal expression, which requires a simple representation of the production plants. In this context, we consider that:

– the speed control of each unit i is reduced to a droop with value δ_i. This is the equivalent to say that the variation in power requested from the unit is instantaneous and proportional to the speed variation to be corrected ($\Delta P_{mi\_ref} = -\frac{1}{\delta_i}\Delta f$);

– the system {actuator + process} of all the plants in the system tests the same transfer function, where the constants T_1 and T_2 are expressed in seconds:

$$\frac{\Delta P_{mi}(s)}{\Delta P_{mi\_ref}(s)} = \frac{1+T_1 s}{1+T_2 s} \quad [3.17]$$

For all of the n plants $G_1 ... G_n$ connected to the grid, of which n' plants participate in primary control according to droops $\delta_1 ... \delta_{n'}$, we will use δ_{eq} to write the equivalent droop of the system in *VAbase*. Its calculation is carried out in a stabilized system, where we write, on the one hand:

$$\Delta P_{m\_ref} = -\frac{1}{\delta_{eq}}\Delta f \quad [3.18]$$

And, on the other hand:

$$\Delta P_{m\_ref} = \sum_{i=1}^{n} \Delta P_{mi\_ref} = \sum_{i=1}^{n'} \Delta P_{mi\_ref} = -\sum_{i=1}^{n'} \left(\frac{1}{\delta_i}\Delta f \right) \quad [3.19]$$

From this, we deduce the following expression and the complete schema of the simplified dynamic model of an electrical system around a point of equilibrium shown in Figure 3.31:

$$\delta_{eq} = \left(\sum_{i=1}^{n'} \frac{1}{\delta_i} \right)^{-1} \quad [3.20]$$

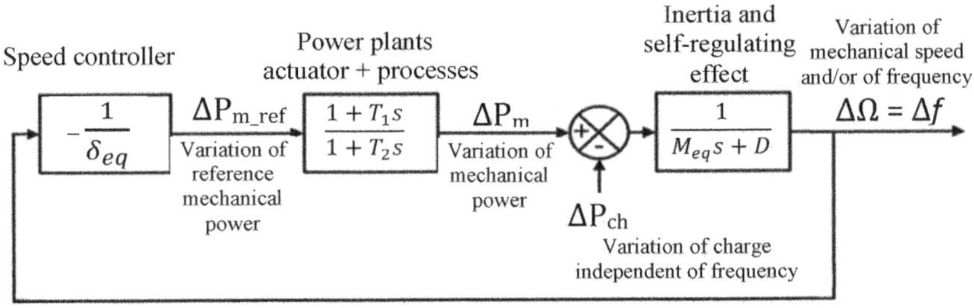

Figure 3.31. *Simplified model of an electrical system whose primary control is provided by production plants using identical technology [DEL 10]*

3.5.5.2. *Temporal frequency evolution*

On the basis of Figure 3.31 thus obtained, we express the frequency transient in the form of the following transfer function:

$$\frac{\Delta f(s)}{\Delta P_{ch}(s)} = -\frac{\dfrac{\delta_{eq}}{\delta_{eq}D+1}(T_2 s+1)}{\dfrac{\delta_{eq}T_2 M_{eq}}{\delta_{eq}D+1}s^2 + \dfrac{\delta_{eq}M_{eq} + \delta_{eq}T_2 D + T_1}{\delta_{eq}D+1}s+1} \quad [3.21]$$

To reduce the amount of writing that follows, we define the constants N_{eq}, D_{1eq} and D_{2eq}, which are characteristics proper to the electrical system being studied:

$$N_{eq} = \frac{\delta_{eq}}{\delta_{eq}D+1} \quad [3.22]$$

$$D_{1eq} = N_{eq}T_2 M_{eq} \quad [3.23]$$

$$D_{2eq} = \frac{\delta_{eq} M_{eq} + \delta_{eq} T_2 D + T_1}{\delta_{eq} D + 1} \qquad [3.24]$$

We consider that the stress imposed on the system is a level of charge of amplitude ΔP_{ch0} ($\Delta P_{ch}(s) = \frac{\Delta P_{ch0}}{s}$), which gives, once all calculations have been made:

$$\Delta f(s) = -\Delta P_{ch0} N_{eq} \left[\frac{T_2}{D_{1eq} s^2 + D_{2eq} s + 1} + \frac{1}{s\left(D_{1eq} s^2 + D_{2eq} s + 1\right)} \right] \qquad [3.25]$$

The reverse Laplace transform is used to reach the temporal expression of the frequency variation after the incident which, in the case of a negative determinant $D_{2eq}^2 - 4D_{1eq}$, which interests us in practice, is written as, defining $\theta = \arctan\left(\frac{\xi}{\sqrt{1-\xi^2}}\right)$:

$$\Delta f(t) = -\Delta P_{ch0} N_{eq} \left[1 + \frac{e^{-\xi \omega_n t}}{\sqrt{1-\xi^2}} \left(T_2 \omega_n \sin\left(\sqrt{1-\xi^2}\,\omega_n t\right) - \sin\left(\sqrt{1-\xi^2}\,\omega_n t + \theta\right) \right) \right] \qquad [3.26]$$

3.5.5.3. *Determination of minimal frequency*

The expression above is first derived to express the time $t_{\min}$ at the end of which it cancels itself, that is, the time at the end of which the minimum frequency is obtained on the system after the beginning of the disequilibrium event.

With all calculations made, we get:

$$t_{\min} = \frac{\frac{\pi}{2} + \xi}{\omega_n \sqrt{1-\xi^2}} \quad \text{with } \tan\xi = \frac{1 - T_2 \omega_n \cos\theta}{T_2 \omega_n \sin\theta} \qquad [3.27]$$

This time depends on the parameters of the system (T_1, T_2, δ_{eq}, D and M_{eq} at the origin of ω_n, ξ, θ and ζ) and is independent of the depth of the imbalance. Finally, we reach the literal expression of the value of the frequency dip obtained following the disequilibrium event of amplitude ΔP_{ch0}:

$$\Delta f_{\max} = -\Delta P_{ch0} N_{eq} \left[1 + \frac{e^{-\xi\omega_n t_{\min}}}{\sqrt{1-\xi^2}} \left(T_2 \omega_n \sin\left(\sqrt{1-\xi^2}\,\omega_n t_{\min}\right) - \sin\left(\sqrt{1-\xi^2}\,\omega_n t_{\min} + \theta\right)\right)\right] \quad [3.28]$$

3.5.5.4. *Sizing of storage*

The results established above show that the depth $\Delta f_{\max}$ reached during the incident, which is expressed here in hertz, is proportional to the amplitude ΔP_{ch0} of the production–consumption imbalance:

$$\Delta f_{\max} = f_0 \Delta f_{\max}^{pu} = -f_0 (\Delta P_{ch0}^{pu} \times C_{transitory\_f}^{pu}) = -\Delta P_{ch0} \times C_{transitory\_f}^{MW/Hz} \quad [3.29]$$

In this expression, the term $C_{transitory\_f}^{MW/Hz}$ (Hz/MW]) characterizes the dynamic behavior of the electrical system modeled, whose expression can be easily identified using equation [3.28] above. This data depend, for a given operating point, on five parameters: the kinetic energy of the rotating masses (or the equivalent system launch time M_{eq}), the primary controlling energy (or the equivalent system droop δ_{eq}), the self-control of the charge (whose amplitude is characterized by the coefficient D) and the time constants of the simplified dynamic model of the production groups (T_1 and T_2 of the lead and lag filter). For the calculations, the response time of the storage system is disregarded in the face of the other phenomena in play on the scale of the electromechanical transient; in other words, at first approximation, the impulse reserve $P_{storage}$ is released instantly and immediately after the incident in order to reduce the imbalance between production and consumption at a value:

$$\Delta P_{ch0}' = \Delta P_{ch0} - P_{storage} \quad [3.30]$$

The sizing of the storage system is a matter of estimating the storage power required to achieve an objective Δf_{max_target} depth of the frequency dip permissible following the imbalance ΔP_{ch0}. Using equation [3.29], we write:

$$P_{storage} = \frac{\Delta f_{\text{max\_target}}}{C_{transitory\_f}^{\text{MW/Hz}}} + \Delta P_{ch0} \quad [3.31]$$

According to the calculations made on the basis of the data available to us, 3–7 MW installed in Guadeloupe make it possible to keep the frequency higher than 49 Hz under the hypotheses considered. These results give an initial value but must be used carefully, given the various simplifications necessary in order for the proposed approach to yield usable analytical expressions. The principal limits of the model are related to the fact that nonlinearities of the production groups are not taken into account.

3.5.6. *Stage 2: dynamic simulations*

To go further, a detailed characterization of the proposed application and the updating of control laws have then been carried out via a series of simulations with the software Eurostag v4.4. To do this, the dynamic representation of a storage unit based on a supercapacitors bank has been introduced by sets of 500 kW into the model of the Guadeloupe grid for the three operating scenarios. Based on the dynamic support specifications given in section 3.5.3, a local supervision strategy including various operating modes has been developed, and its performances have been studied in detail.

In the end, it appears that the application and controls proposed make it possible to take advantage of the very short response times of modern storage technologies for the benefit of power system security and reliability. The rapid release of an impulse reserve following an incident reduces the temporary overloads imposed on groups remaining connected to the grid. Consequently, this dynamic support moderates frequency differences, with a favorable impact that may prove notable as part of the efforts of islanded grid operators to improve the continuity of their power supply. For example, for scenario 1, 3.5 MW of rapid storage assist the release of the primary reserve enough to avoid consumer shedding; in Figure 3.32, in which the automated shedding process is deactivated in order to facilitate comparisons, the corresponding curve remains above the initial threshold of 48.5 Hz.

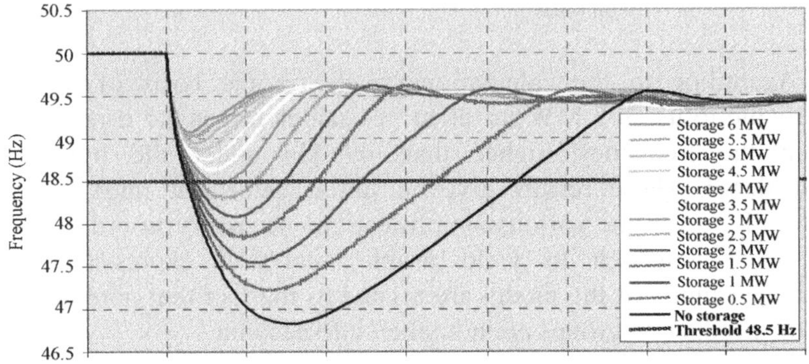

Figure 3.32. *Evolution of frequency in dynamic model of the Guadeloupe grid following the loss of a generation plant, scenario 1 of [DEL 10] with storage. For a color version of the figure, see www.iste.co.uk/robyns/powergrids.zip*

Various points have been discussed to complete this characterization, such as recharge management to guarantee the availability of the service proposed, the choice of operating voltage of supercondensers and the behavior of storage systems in the event of HV voltage drops. Finally, via scenarios 2 and 3, it has been shown that dynamic support of frequency control can also be of interest in the presence of high penetration rates of decentralized production using a renewable source [DEL 12].

3.5.7. *Stage 3: experimental laboratory implementation*

As further support, the models used have been refined and validated through experimental tests to the L2EP real-time simulation platform. As shown in Figure 3.33, a low-power supercapacitor unit (42 F, 194 V, 10 kVA) has been interfaced with the simulated Guadeloupe grid in order to test its control in operating conditions similar to reality.

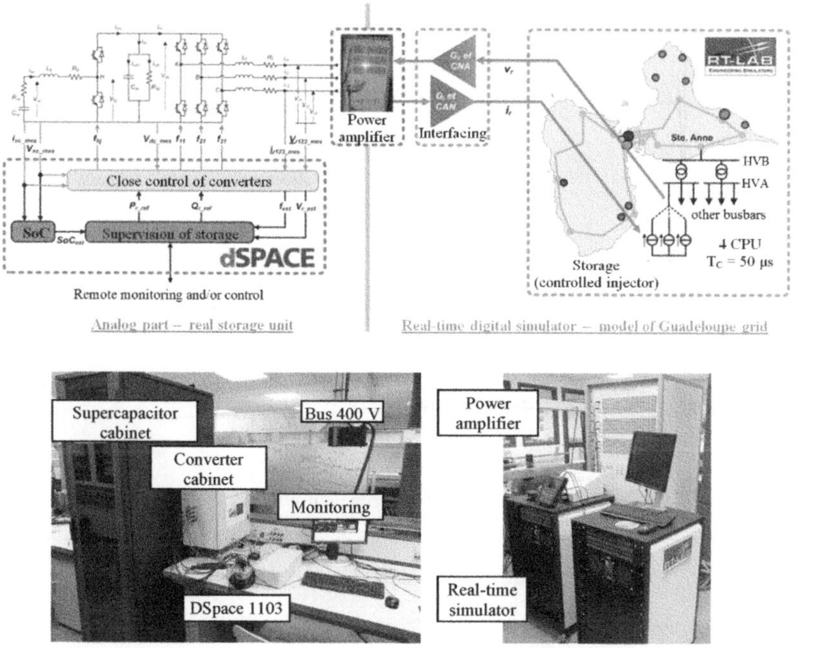

Figure 3.33. *Experimental setup for testing dynamic support [DEL 10]. For a color version of the figure, see www.iste.co.uk/robyns/powergrids.zip*

Figure 3.34 presents an example of the results of this experimental application of dynamic support to a laboratory prototype. Initially, the storage system is on standby and charged at V_{sc_ref} = 160 V (a). The triggering of a large generation plant is controlled on the real-time simulator and causes a rapid frequency drop (b) to which the supervision system of the real storage unit reacts. Initially, full power is rapidly released in order to avoid a shedding of consumption. The support is then prolonged (d) until the minimum authorized state of charge. Between t_0 + 2 and t_0 + 4 min, the reconnection and progressive increase in power of the generation plant takes the system frequency back to 50 Hz. In the absence of any new disturbance for a predetermined waiting period, the storage system is recharged at reduced power from t_0 + 9 min (f) until the reference voltage V_{sc_ref} (g) is obtained.

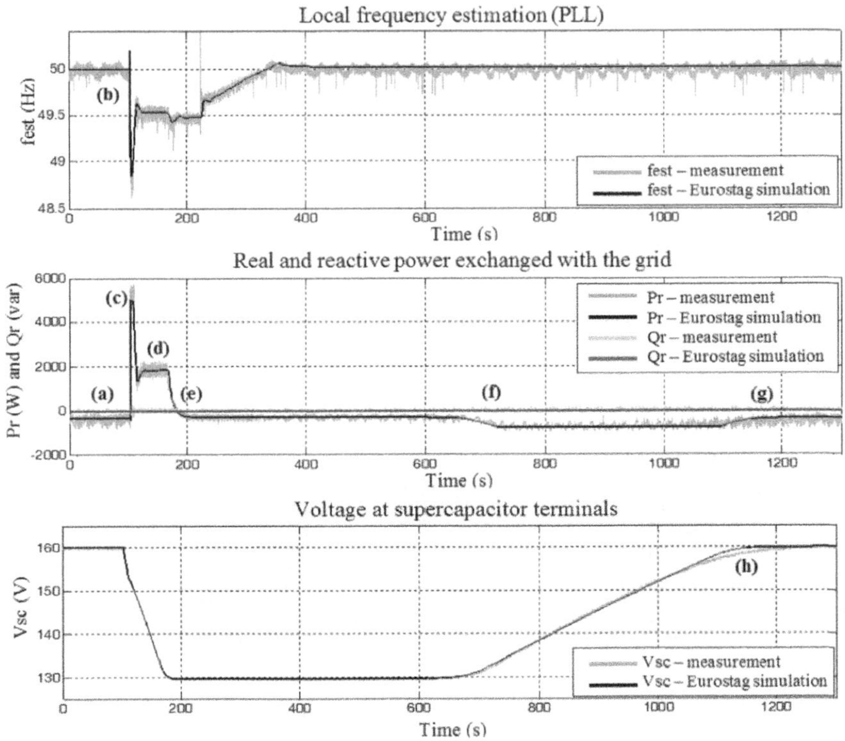

Figure 3.34. *Response of laboratory prototype to a frequency variation and comparison with simulation results [DEL 10]. For a color version of the figure, see www.iste.co.uk/robyns/powergrids.zip*

This series of tests has been used to validate the models used for studies and to improve supervision schemas considerably. The control modifications required to ensure correct operating of the real storage unit were then integrated into the dynamic simulations in order to refine the characterization of the service being studied.

3.5.8. *Economic value making*

Based on the sizing carried out, elements of economic value making were also established for the service. According to the data on which the calculations were based, several megawatts of impulse reserve could avoid up to 50% of sheddings on an electrical system such as the one on the island of Guadeloupe. This reduction of outages, when translated into economic terms using standard ideas of power interruptions and NDE, in some cases exceeds the present investment and exploitation costs for ultracapacitor-based storage units. The value resulting from dynamic support might also aid in increasing the profitability of systems dedicated mainly to other applications.

3.5.9. *Conclusion*

This case study has, in addition to supplying details regarding an innovative service, shown examples of the methods and tools that can be used to study the contributions of storage to the improved operation of electrical systems.

3.6. General conclusion

The first part of this chapter specified the basic operating characteristics of electrical systems. Next, the services able to be provided by a storage unit for various stakeholders in these systems were presented. To complete these elements and to shed light on study methods, the final sections were dedicated to targeted examples of the contributions of storage to the treatment of congestion events and the dynamic support of frequency control in the event of sudden imbalances in an islanded grid.

Due to the investment costs of storage systems, which remain high, the services possible for a stakeholder are not enough to justify this solution economically in the face of other conventional or innovative solutions. One possible way to maximize the benefits generated by a storage system is to mutualize a certain number of services on behalf of its owner, or for other stakeholders against fees [DEL 09]. The decision to invest must logically be based on a main service. In order to maximize the value of a storage system, additional services can be added to it, subject to a validation of technological feasibility. Once a cohesive list of services has been established, a business model must be constructed, identifying the relevant stakeholders and the one to whom the major part of the value would be allocated. Multiservice-type operation involves a specific request of the storage system; the system operator must integrate smart decision algorithms in order to select different services according to need. Depending on the case, command and control can be carried out exclusively in real time, or they may require a degree of upstream planning based, for example, on predictions of renewable production.

3.7. Bibliography

[ABO 05] ABOU-ÇHACRA F., Valorisation et optimisation du stockage d'énergie dans un réseau d'énergie électrique, Thesis, University of Paris XI, UFR Scientifique d'Orsay, 2005.

[ACK 01] ACKERMANN T., ANDERSSON G., SODER L., "Distributed generation: a definition", *Electric Power Systems Research*, vol. 57, pp. 195–204, 2001.

[ACK 05] ACKERMANN T., *Wind Power in Power Systems*, Wiley, 2005.

[AES 14] AES, available at www.aesenergystorage.com, 2014.

[BAR 04] BARTON J., INFIELD D., "Energy storage and its use with intermittent renewable energy", *IEEE Transaction on Energy Conversion*, vol. 19, no. 2, pp. 441–448, June 2004.

[BOU 09] BOUHALI O., FRANCOIS B., BERKOUK E. *et al.*, "Power sizing and control of a three-level NPC converter for grid connection of wind generators", *Electromotion Journal,* vol. 16, no. 1, pp. 38–49, March 2009.

[CRA 03] CRAPPE M., *Commande et régulation des réseaux électriques*, Hermès-Lavoisier, Paris, 2003.

[DEL 09] DELILLE G., FRANCOIS B., MALARANGE G., "Construction d'une offre de services du stockage pour les réseaux de distribution dans un contexte réglementaire dérégulé", *European Journal of Electrical Engineering*, vol. 12, nos. 5–6, pp. 733–762, 2009.

[DEL 10] DELILLE G., Contribution du Stockage à la Gestion Avancée des Systèmes Electriques, Approches Organisationnelles et Technico-économiques dans les Réseaux de Distribution, Thesis, l'Ecole Centrale de Lille, L2EP, 2010.

[DEL 12] DELILLE G., FRANCOIS B., MALARANGE G., "Dynamic frequency control support by energy storage to reduce the impact of wind and solar generation on isolated power system's inertia", *IEEE Transactions on Sustainable Energy,* vol. 3, no. 4, pp. 931–939, 2012.

[DEV 09] DEVATINE L., "Postes à haute et très haute tension: Rôle et structure", *Techniques de l'Ingénieur*, reference D4570, 2009.

[DIR 09] DIRECTIVE 2009/28/CE relative à la promotion de l'utilisation de l'énergie produite à partir de sources renouvelables et modifiant puis abrogeant les directives 2001/77/CE et 2003/30/CE, 2009.

[DOU 02] DOULET A., "Le concept d'énergie non distribuée, outil d'aide à la décision dans la gestion des réseaux électriques", *Réalités industrielles*, pp. 62–68, 2002.

[EIR 09] EIRGRID, EirGrid Grid Code, Version 3.3, [Online], 30 January 2009, available at http://www.eirgrid.com.

[EN 99] EN 50160, Voltage characteristics of electricity supplied by public distribution systems, 1999.

[ENT 14] ENTSOE, *Continental Europe Operation Handbook,* [Online], 2014, available at https://www.entsoe.eu/publications/system-operations-reports/operation-handbook/Pages/ default.aspx.

[EON 05] EON, Wind report, [Online], 2005, available at http://www.transpower.de.

[EPR 02] EPRI (Electric Power Research Institute), Primer on distributed energy resources for distribution planning, Palo Alto, CA, 1004644, 2002.

[EPR 03] EPRI (Electric Power Research Institute), Handbook of energy storage for transmission and distribution applications, Technical Report 1001834, Palo Alto, CA, and the U.S. Department of Energy, Washington, DC, December 2003.

[EPR 05a] EPRI (Electric Power Research Institute), VRB energy storage for voltage stabilization: testing and evaluation of the Pacificorp Vanadium Battery Energy Storage System at Castle Valley, Palo Alto, CA, 1008434, 2005.

[EPR 05b] EPRI (Electric Power Research Institute), Wind power integration: energy storage for firming and shaping, Palo Alto, CA, 1008388, 2005.

[ERD 08] ERDF, Description physique du réseau de distribution, ERDF-NOI-RES_07E, 2008.

[EYE 04] EYER J., IANNUCCI J., COREY G., Energy storage benefits and market analysis handbook – a study for the DOE Energy Storage Program, Sandia National Laboratories, Technical Report SAND2004-6177, December 2004.

[EYE 05] EYER J., IANNUCCI J., BUTLER P., Estimating electricity storage power rating and discharge duration for utility transmission and distribution deferral – a study for the DOE Energy Storage Program, Sandia National Laboratories, Technical Report SAND2005-7069, November 2005.

[EYE 10] EYER J., COREY G., Energy storage for the electricity grid: benefits and market potential assessment guide – a study for the DOE Energy Storage Program, Sandia National Laboratories, Technical Report SAND10-0815, February 2010.

[FRI 08] FRIDLEIFSSON I., BERTANI R., LUND J. *et al.*, "The possible role and contribution of geothermal energy to the mitigation of climate change", *IPCC Proceeding*, 20–25 June 2008.

[GAU 97] GAUDRY M., BOUSQUE J.-L., *Lignes aériennes: échauffements et efforts électrodynamiques*, Technique de l'Ingénieur, vol. D4439, February 1997.

[GAU 05] GAUTIER E., BOUSSEAU P., BELHOMME R. *et al.*, "Revue de solutions pour l'intégration de l'éolien dans les réseaux électriques", *REE*, no. 5, pp. 59–64, May 2005.

[GER 08] GERMAN BUNDESTAG, Update of feed in tariff for renewable energies in Germany, Feed in International Corporation, 2008.

[HAR 05] HARMAND Y., NEBAS-HAMOUDIA C., LARRIPA B. *et al.*, "Le mécanisme d'ajustement: Comment assurer l'équilibre production-consommation dans un marché ouvert à la concurrence?", *REE*, vol. 6/7, pp. 91–104, July 2005.

[HOL 04] HOLTTINEN H., The impact of large scale wind power production on the Nordic electrical system, Thesis, Helsinki University of Technology, Espoo, Finland, 2004.

[IAN 05] IANNUCCI J., EYER J., ERDMAN B., Innovative applications of energy storage in a restructured electricity marketplace phase III final report – a study for the DOE Energy Storage Program, Sandia National Laboratories, Technical Report SAND2003-2546, March 2005.

[KAN 14] KANCHEV H., Gestion des flux énergétiques dans un système hybride de sources d'énergie renouvelable: Optimisation de la planification opérationnelle et ajustement d'un micro réseau électrique urbain, Thesis, l'Ecole Centrale de Lille et de l'Université Technique de Sofia, 24 January 2014.

[KAS 07] KASHEM M., LEDWICH G., "Energy requirement for distributed energy resources with battery energy storage for voltage support in three-phase distribution lines", *Electric Power Systems Research*, vol. 77, no. 1, pp. 10–23, 2007.

[KOE 08] KOEPPEL G., KORPAS M., "Improving the network infeed accuracy of non-dispatchable generators with energy storage devices", *Electric Power Systems Research*, vol. 78, pp. 2024–2036, 2008.

[KUN 86] KUNISCH H.J., KRAMER K.G., DOMINIK H., "Battery energy storage – another option for load-frequency-control and instantaneous reserve", *IEEE Transactions on Energy Conversion*, vol. EC-1, no. 3, September 1986.

[KUN 94] KUNDUR P., *Power System Stability and Control*, McGraw-Hill, New York, 1994.

[LU 10a] LU D., FAKHAM H., ZHOU T. *et al.*, "Application of Petri Nets for the energy management of a photovoltaic based power station including storage units", *Renewable Energy*, Elsevier, vol. 35, no. 6, pp. 1117–1124, 2010.

[LU 10b] LU D., Conception et contrôle d'un générateur PV actif à stockage intégré: Application à l'agrégation de producteurs-consommateurs dans le cadre d'un micro réseau intelligent urbain, thèse de doctorat de l'Ecole Centrale de Lille, 16 December 2010.

[MAR 98] MARQUET A., LEVILLAIN C., DAVRIU A. *et al.*, *Stockage d'électricité dans les systèmes électriques*, Techniques de l'Ingénieur, reference D4030, 1998.

[MOR 07] MOREIRA C., RESENDE O., PEÇAS LOPES J., "Using low voltage microgrids for service restoration", *IEEE Transactions on Power Systems*, vol. 22, no. 1, pp. 395–403, 2007.

[MUL 03] MULTON B., *Production d'énergie électrique par sources renouvelables*, Techniques de l'Ingénieur, traité Génie Electrique D 4005, 2003.

[NOR 07] NORRIS B., NEWMILLER J., PEEK G., NAS battery demonstration at American Electric Power – a study for the DOE Energy Storage Program, Sandia National Laboratories, Technical Report SAND2006-6740, 2007.

[NOU 07] NOURAI A., Installation of the first distributed energy storage system (DESS) at American Electric Power (AEP), Sandia National Laboratories, Technical Report SAND2007-3580, 2007.

[NOU 08] NOURAI A., KOGAN V.I., SCHAFER C.M., "Load leveling reduces T&D line losses", *IEEE Transactions on Power Delivery*, vol. 23, no. 4, pp. 2168–2173, 2008.

[NOU 10] NOURAI A., KEARNS D., "Batteries included", *IEEE Power and Energy Magazine*, vol. 8, no. 2, pp. 49–54, 2010.

[OUD 06] OUDALOV A., CHARTOUNI D., OHLER C. *et al.*, "Value analysis of battery energy storage applications in power systems", *Proceedings of the IEEE PES Power System Conference and Exposition (PSCE06)*, Atlanta, pp. 2206–2211, 2006.

[RIC 06] RICHARDO O., Réglage coordonné de tension dans le réseaux de distribution à l'aide de la production décentralisée, PhD Thesis, INP Grenoble, 2006.

[RIF 09] RIFFONNEAU Y., Gestion des flux énergétiques dans un système photovoltaïque avec stockage connecté au réseau, Thesis, Joseph Fourier University, 23 October 2009.

[ROB 04] ROBYNS B., BASTARD P., "Production décentralisée d'électricité: contexte et enjeux techniques", *Revue 3EI*, no. 39, pp. 5–13, December 2004.

[ROB 06] ROBYNS B., DAVIGNY A., SAUDEMONT C. *et al.*, "Impact de l'éolien sur le réseau de transport et la qualité de l'énergie", *Journal sur l'enseignement des sciences et technologies de l'information et des systèmes, J3eA*, vol. 5, no. 1, 2006.

[ROB 12] ROBYNS B., DAVIGNY A., BRUNO F. *et al.*, *Production d'énergie électrique à partir des sources renouvelables*, Hermès-Lavoisier, Paris, 2012.

[RTE 04] RTE, "Mémento de la sûreté du système électrique", 2004.

[RTE 14] RTE, Référentiel Technique de RTE, [Online], 2014, available at http://www.rte-france.com.

[RUE 12] RUESTER S., VASCONCELOS J., HE X. *et al.*, Electricity storage: how to facilitate its deployment and operation, European University Institute, Technical Report, 2012.

[RWE 09] RWE, Didcot Power Stations, [Online], 2009, available at http://www.npower.com.

[SAB 06] SABONNADIERE J.-C., *Nouvelle technologies de l'énergie 1: Les énergies renouvelables*, Hermès-Lavoisier, Paris, 2006.

[SAB 07a] SABONNADIERE J.-C., HADJSAID N., *Lignes et réseaux électriques 1: lignes d'énergie électrique*, Hermès-Lavoisier, Paris, 2007.

[SAB 07b] SABONNADIERE J.-C., HADJSAID N., *Lignes et réseaux électriques 2: méthodes d'analyse des réseaux électriques*, Hermès-Lavoisier, Paris, 2007.

[SAB 08a] SABONNADIERE J.-C., HADJSAID N., *Lignes et réseaux électriques 3: la libéralisation des marchés*, Hermès-Lavoisier, Paris, 2008.

[SAB 08b] SABONNADIERE J.-C., HADJSAID N., *Lignes et réseaux électriques 4: exercices et problèmes,* Hermès-Lavoisier, Paris, 2008.

[UCT 04] UCTE, Final report of the Investigation Committee on the 28 September 2003 blackout in Italy, April, [Online], 2004, available at https://www.entsoe.eu.

[UCT 07] UCTE, Final report on the disturbances of 4 November 2006, [Online], 2007, available at https://www.entsoe.eu.

[UK 14] UK POWER NETWORKS, "Demonstrating the benefits of short-term discharge energy storage on an 11kV distribution networks", version 1.1, June 2014.

[VER 09] VERGNOL A., SPROOTEN J., ROBYNS B. *et al.*, "Gestion des congestions dans un réseau intégrant de l'énergie éolienne", *Revue 3EI*, no. 59, pp. 63–72, 2009.

[VER 11a] VERGNOL A., SPROOTEN J., ROBYNS B. *et al.*, "Line overload alleviation through corrective control in presence of wind energy", *Electric Power Systems Research*, Elsevier, vol. 81, pp. 1583–1591, July 2011.

[VER 11b] VERGNOL A., ROBYNS B., "Localization of storage by identification of weakness of power systems", *MixGenera 2011*, Leganés, Madrid, Spain, November 2011.

4

Introduction to Fuzzy Logic and Application to the Management of Kinetic Energy Storage in a Hybrid Wind-Diesel System

4.1. Introduction

In section 4.2 of this chapter, the basic concepts of fuzzy logic will be introduced. They will then be applied to the management of a kinetic energy storage system that is part of a hybrid wind-diesel generator system powering an isolated site such as an island.

This storage system is composed of an asynchronous machine driving a flywheel, which must ensure the smoothing of the total power generated subject to variations in wind and load consumption. The objective of the fuzzy logic supervisor is to smooth the power generated by windmills, which is by nature highly variable [ROB 12b]. The performances of the proposed fuzzy logic supervisor will be compared with simplified management based on a low-pass filter, filtering the power generated by the windmill, a comparison that will illustrate the interest of taking the state of charge of the storage system into account in energy management.

4.2. Introduction to fuzzy logic

The goal of fuzzy logic is to translate linguistic rules into a mathematical form, known as fuzzy rules, which describe the observations and reactions of a human operator controlling a process. The results deduced from fuzzy logic are utterly deterministic. The concept of fuzziness is to be considered in relation to the concept of uncertainty present in most systems with which we interact in practice; this uncertainty is formalized mathematically in the form of a membership function.

4.2.1. *Principle of fuzzy reasoning*

The driver of a car approaching a traffic light will have a reaction adapted to various situations that can be expressed as follows:

– if the light is *red* and if my speed is *high* and if the light is *near*, then I brake *hard*;

– if the light is *red* and if my speed is *low* and if the light is *far away*, then I maintain my speed;

– if the light is *yellow* and if my speed is *medium* and if the light is *far away*, then I brake *gently*;

– if the light is *green* and if my speed is *low* and if the light is *near*, then I accelerate.

This example does not exhaustively cover all of the possible situations. It shows that we can naturally express our actions and reactions by means of fuzzy logic. The concepts "high", "near", "hard", "low", "far", "medium" and "gently" are fuzzy in the sense that we do not assign a precise numerical value to them in our reasoning. We often use these concepts in everyday life to describe situations or to carry out actions. If we measured the values of the variables in play at the time of the action, however, we would obtain precise numerical values. For example, the first rule given above would be expressed as follows:

– if the light is red and if my speed is higher than *85.6 km/h* and if the light is at least *62.3 m* away, then I press on the brake pedal with a force of *33.2 N*.

You might say that our brain functions using fuzzy logic that understands input variables in an approximate way (low, high, far away and near), does the same for output variables (gentle or hard braking) and establishes a set of rules used to determine the outputs as a function of the inputs. Note that to drive a car we do not need to know a model of the car, which would be very complex; rather, we use our knowledge of driving and our vehicle. This is the extremely powerful approach, a part of human nature, which we are attempting to formalize so that we can reproduce it for the management of complex systems.

4.2.2. *Fuzzy logic and Boolean logic*

Fuzzy logic is based on two main concepts:

– fuzzy sets and variables and the associated operators;

– decision-making based on the basic "IF-THEN" rule, which we call fuzzy inference.

Fuzzy sets and the associated operators were defined by Professor Lofti Zadeh of the University of Berkeley in 1965.

Based on the classic theory of sets, in the case of Figure 4.1, we can write (with set U being the universe of discourse of variable x and A being a subset of U):

– if μ_A is the membership function of set A:

$$\begin{aligned} \forall x \in U \quad & \mu_A(x) = 0 \quad \textit{if } x \notin A \\ & \mu_A(x) = 1 \quad \textit{if } x \in A \end{aligned}$$

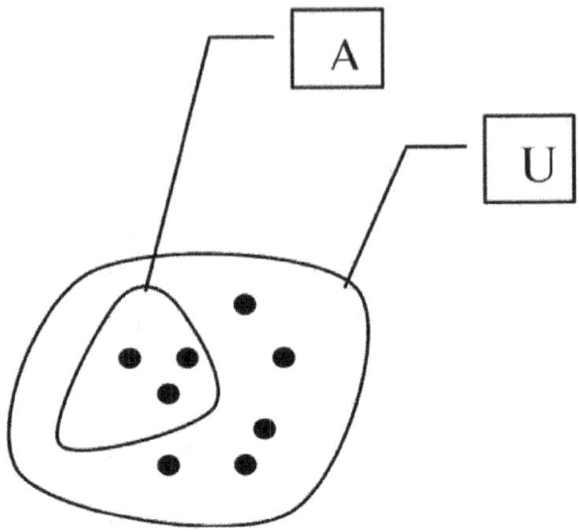

Figure 4.1. *Set and subset*

In fuzzy logic, the membership of x in subset A is not binary as in Boolean logic; rather, this belonging is explained by a degree of membership as follows:

– if μ_A is the membership function of set A:

$$\forall x \in U \quad \mu_A(x) \in [0;1]$$

If $\mu_A(x) = 0.30$, x belongs to fuzzy set A with a degree of membership of 30%. A fuzzy set is completely determined by its membership function.

As an example, Figure 4.2 illustrates, for a “height of a person” variable, the subsets “small”, “medium” and “big”. We can deduce from Figure 4.2(a) that membership in the “small” subset, of a person less than 1.6 m tall, is 1; of a person with a height of 1.65 m is 0.5; and of a person taller than 1.7 m is 0. The concept of fuzziness appears for heights of between 1.6 and 1.7 m. Usually, the three subsets are represented in a single graphic (Figure 4.3) called “membership functions”.

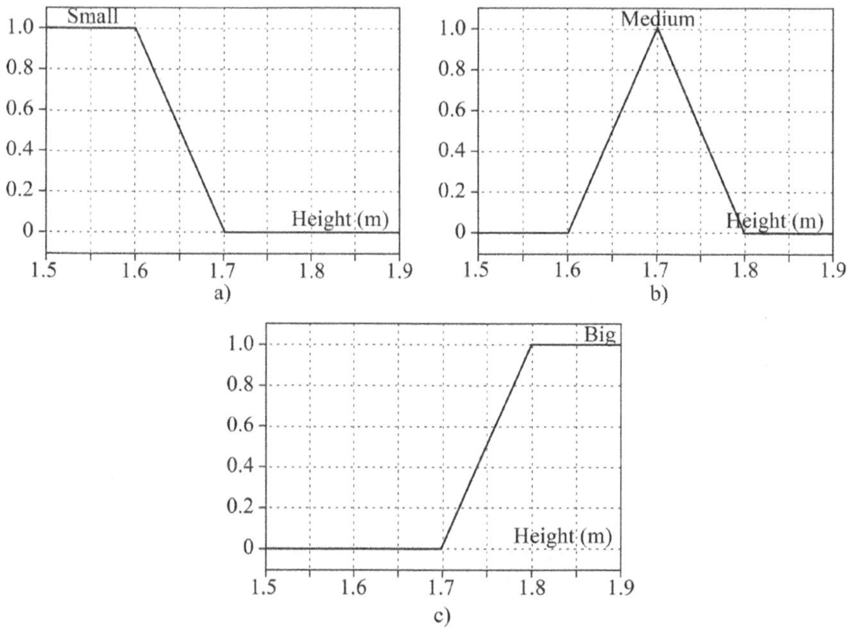

Figure 4.2. *Small a), medium b) and big c) fuzzy sets*

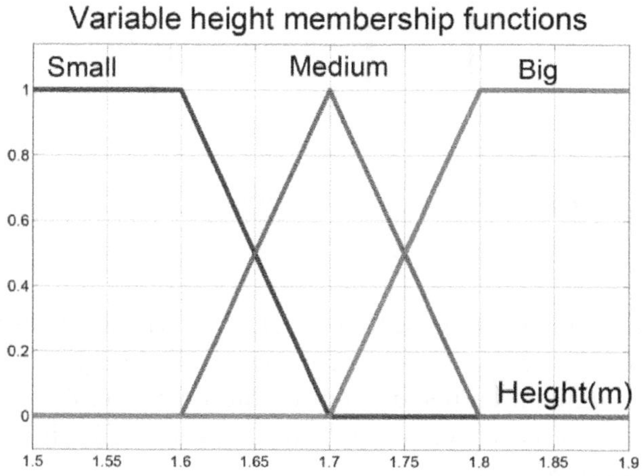

Figure 4.3. *Membership functions of height variable. For a color version of the figure, see www.iste.co.uk/robyns/powergrids.zip*

If Pierre's height is 1.625 m, according to Figure 4.3, we deduce the degree of membership in the different fuzzy subsets:

– "Pierre is small" at a degree of 75%;

– "Pierre is medium" at a degree of 25%;

– "Pierre is big" at a degree of 0%.

In fuzzy logic, the "height of a person" variable is called a *linguistic value*, and the subsets "small", "medium" and "big" are called *linguistic variables*. The height variation of between 1.5 and 1.9 m being considered is called the *universe of discourse*.

In fuzzy logic, we find specific membership functions. If the "medium" subset represented in Figure 4.2(b) is no longer defined by a triangle but rather by a rectangle, we obtain the function shown in Figure 4.4. Since the degree of membership to this subset can only be 0 or 1, we revert to classic Boolean logic.

Figure 4.4. *Membership function in Boolean logic*

In the example given in section 4.2.1, the variable corresponding to the color of the traffic light is composed of singletons. Figure 4.5 shows the singleton corresponding to "the light is red". The value of membership to this subset is 1 only for a given traffic light color, and 0 for all the other colors.

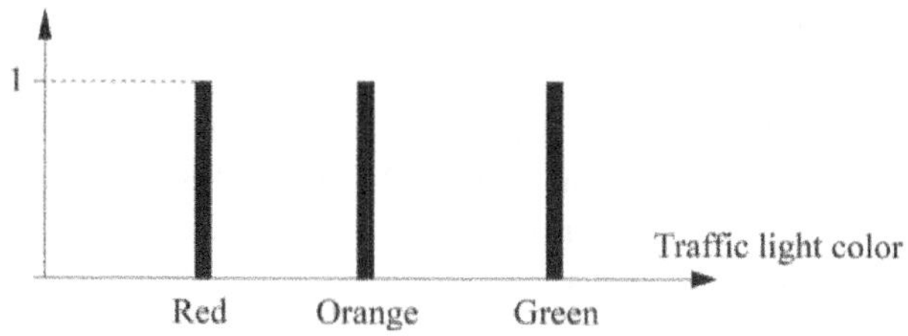

Figure 4.5. *Singleton membership function*

Thus, classic Boolean logic constitutes a particular case of fuzzy logic that is more general. All results obtained in classic Boolean logic must be found using fuzzy logic.

The operators used in fuzzy logic are defined in the same way as in classic logic:

– the AND or Intersection operator can be defined in two ways:

MIN/MAX: $\mu_{A \cap B}(x) = \min(\mu_A(x), \mu_B(x))$ [4.1]

PROD/SUM: $\mu_{A \cap B}(x) = \mu_A(x) \times \mu_B(x)$ [4.2]

– the OR or Union operator can also be defined in two ways:

MIN/MAX: $\mu_{A \cup B}(x) = \max(\mu_A(x), \mu_B(x))$ [4.3]

PROD/SUM: $\mu_{A \cup B}(x) = (\mu_A(x) + \mu_B(x)) / 2$ [4.4]

– finally, the NO or Complement operator is defined as follows:

$\mu_{\overline{A}}(x) = 1 - \mu_A(x)$ [4.5]

4.2.3. *Stages of a fuzzy supervisor*

There are three principal stages in fuzzy reasoning: fuzzification, inference and defuzzification [BUH 94, BOR 98]. These stages are illustrated in Figure 4.6. Input and output variables are usually normalized between –1 and 1.

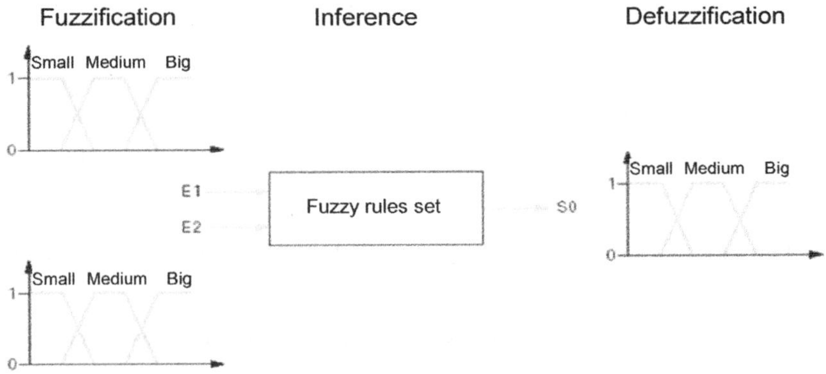

Figure 4.6. *Stages of fuzzy reasoning*

Fuzzy logic systems address process fuzzy input variables and supply results in output variables that are also fuzzy. *Fuzzification* is the stage consisting of the fuzzy quantification of the real values of a variable. This stage has already been illustrated in the example shown in section 4.2.2 and is summarized in Figure 4.7.

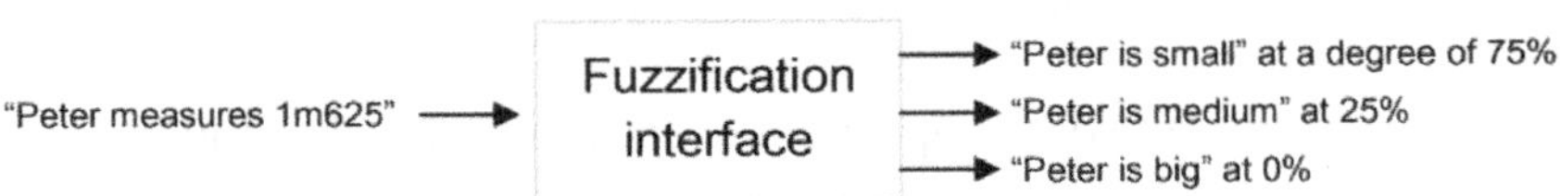

Figure 4.7. *Example of a stage of fuzzification*

The fuzzification of input and output variables is a delicate phase of the process of implementation by fuzzy logic. It is often carried out in an iterative manner and requires experience. Various examples discussed throughout this book will give more specific information about the approach to defining membership functions, and solutions for reducing the empiricism of the approach will be introduced.

For a real variable value, the value of the degree of membership will be a function of:

– the forms chosen for membership functions;

– certain parameters of these membership functions, such as height, core and support.

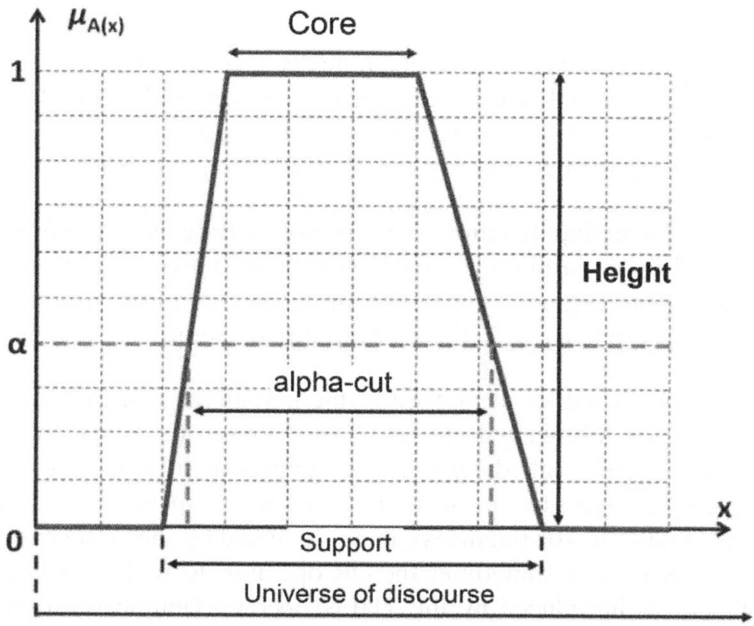

Figure 4.8. *Characteristic values of a fuzzy set*

Consider Figure 4.8, which presents the characterization of a fuzzy set A in universe of discourse X:

– height is the strongest degree of membership with which an element of the universe of discourse belongs to the fuzzy set. If the fuzzy set is normalized, the height will have a value of 1;

– the core is composed of all the elements of the universe of discourse whose membership in the fuzzy set is absolute: {x ε X, such that $\mu_A(x) = \text{height}$};

– the support is composed of all the elements of the universe of discourse for which the membership function is not zero: {x ε X, such that $\mu_A(x) > 0$};

– the alpha-cut α is composed of all the elements of the universe of discourse belonging to a fuzzy set with a degree of membership at least equal to α: {x ε X, such that $\mu_A(x) \geq \alpha$}.

Membership functions can theoretically take any form (Gaussian, sigmoid, trapezoidal, triangular, etc.). However, the triangular and trapezoidal (piecewise linear) forms are the most often used as they facilitate programming and simplify the collection of expertise.

The membership functions we are considering here confirm the concept of fuzzy partition: for each value in the universe of discourse, the sum of the degrees of membership in all the fuzzy ensembles is equal to the height (or 1, in the case of normalization).

The input variables are linked to the output variable by means of rules with the following form: If (X is A) Then (Y is B). In fuzzy logic, this stage is called *inference*. Examples of rules were given in the example discussed in section 4.2.1. When these rules include several conditions (or premises), they are linked by the AND operator (or conjunction), or sometimes the OR operator [ROB 12a], while the conclusion is introduced by the THEN operator (implication). When there are multiple rules, they are linked to each other by the OR operator. The rules in the example in section 4.2.1 are thus expressed in fuzzy logic as follows:

– if the light is red AND if my speed is *high* AND if the light is near THEN I brake *hard*.

OR

– if the light is red AND if my speed is *low* AND if the light is *far away* THEN I maintain my speed.

OR

– if the light is yellow AND if my speed is *medium* AND if the light is *far away* THEN I brake *gently*.

OR

– if the light is green AND if my speed is *low* AND if the light is *near* THEN I accelerate.

OR…

The truer the condition on the input variables, the more the action recommended for the outputs must be respected. The determination of these rules is generally based on the designer's expertise, and is therefore empirical. Beginning in the next chapter, we will introduce a method aimed at aiding in the determination of these laws in order to confine ourselves to the most pertinent ones. Fuzzy algorithms can become complex quickly; therefore, it is important to correctly choose input variables, the number of fuzzy subsets (or linguistic values) and fuzzy laws in order to limit the complexity of the fuzzy algorithm. This point will be discussed throughout this book.

From the inference, it is necessary to determine a deterministic value for the output variables that can be applied to the process. This is the *defuzzification* stage. There are several methods of defuzzification, the most widely used being the determination of the center of gravity of the resulting membership function [BUH 94], which is determined by relationship [4.6]. The denominator represents the surface resulting from the inference stage, while the numerator represents the moment of this surface. Figure 4.9 illustrates the center of gravity for a simple example.

$$Output = \frac{\int_U y\,\mu(y)\,\mathbf{dy}}{\int_U \mu(y)\,\mathbf{dy}} \qquad [4.6]$$

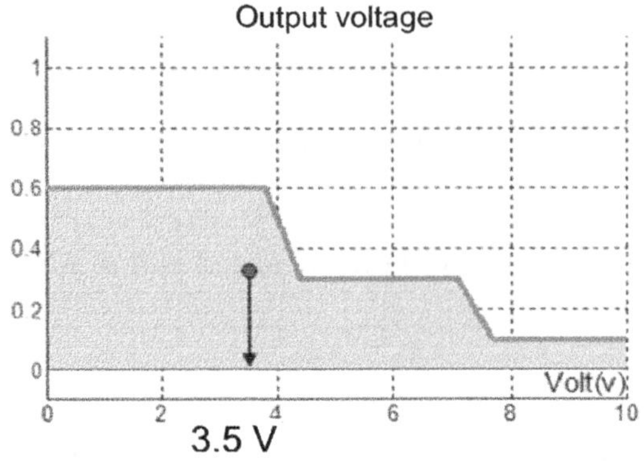

Figure 4.9. *Example of defuzzification stage by center of gravity*

4.2.4. *Example of fuzzy reasoning*

Let us return to the example introduced in section 4.2.1. Two fuzzy inputs must be defined: the speed of the vehicle and the distance in relation to the traffic light. The color of the traffic light is not a fuzzy entry. The membership functions of the speed can be chosen as shown in Figure 4.10.

If the speed of movement is equal to 60 km/h, the membership of this speed value in the fuzzy sets will be as follows:

– membership in the "small" set with a degree of 0;

– membership in the "medium" set with a degree of 0.6;

– membership in the "big" set with a degree of 0.4.

The membership functions of distance in relation to the traffic light can be chosen as shown in Figure 4.11.

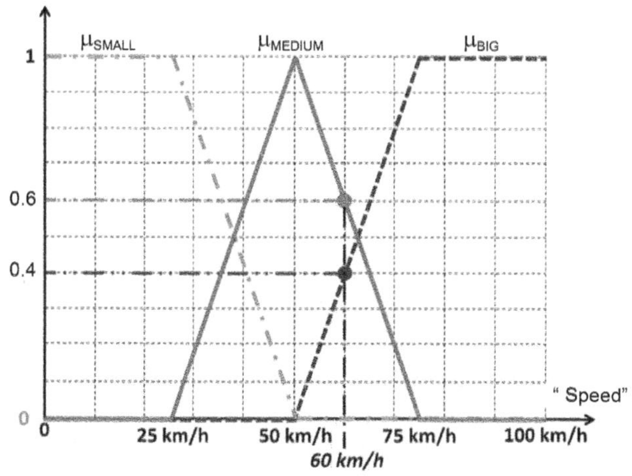

Figure 4.10. *Example of membership functions of speed with three fuzzy sets: "small", "medium" and "big"*

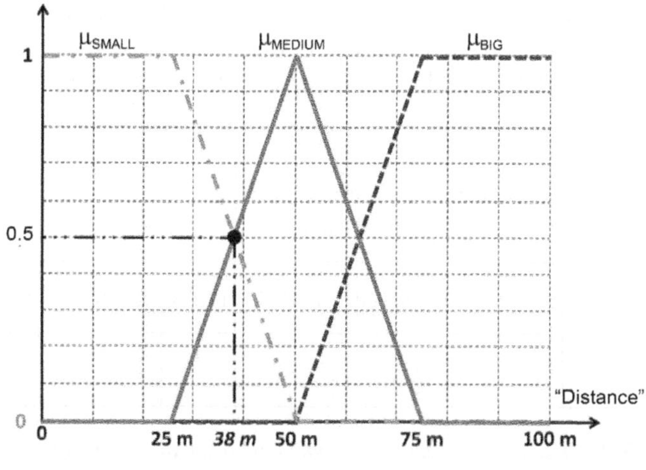

Figure 4.11. *Example of membership functions of distance with three fuzzy sets, "small", "medium" and "big"*

The output variable concerning the intensity of braking is also a fuzzy variable. The membership functions of the braking variable can be chosen as shown in Figure 4.12.

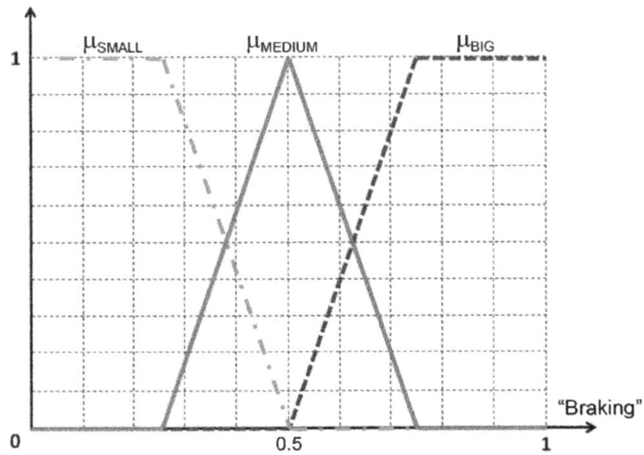

Figure 4.12. *Example of membership functions of the braking output variable with three fuzzy sets, "small", "medium" and "big"*

To illustrate the inference stage, let us consider the two rules from Table 4.1 that are applicable when the traffic light is red, with a "speed" value equal to 60 km/h and a "distance" value of 38 m.

...	...	...	...	...	...	...	...	...	...
R_1	If	Distance	SMALL	And	Speed	BIG	THEN	BRAKING	BIG
R_2	If	Distance	SMALL	And	Speed	MEDIUM	THEN	BRAKING	MEDIUM
...	...	...	...	...	...	...	...	...	...

Table 4.1. *Example of fuzzy rules applicable when the traffic light is red*

The inference mechanism determines the degree of activation of each rule and the implications. Applying the MIN/MAX method to create the AND and OR operators, the illustration in Figure 4.13 shows that these degrees have values of 0.4 and 0.5, respectively, for R_1 and R_2.

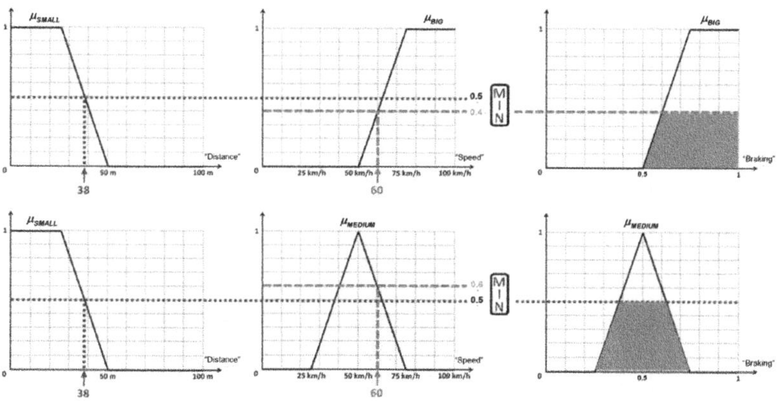

Figure 4.13. *Illustration of inference mechanism. Degrees of activation and conclusions of rules R_1 (top) and R_2 (bottom)*

The final stage of this inference phase uses the max operator (Figure 4.14). The resulting fuzzy set for the "Braking" output variable is obtained by aggregating the previously determined conclusions.

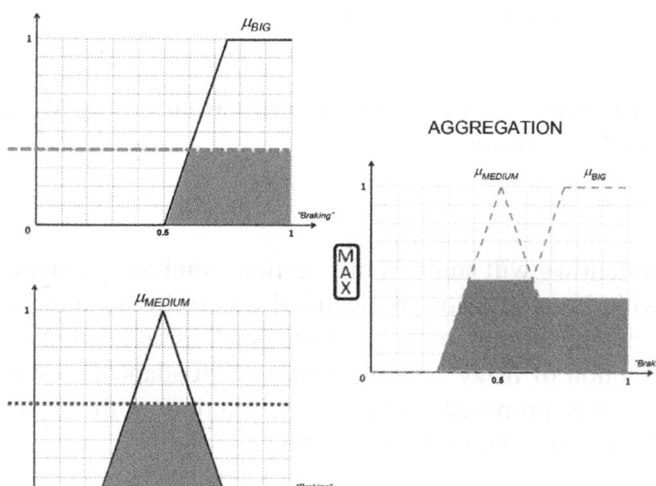

Figure 4.14. *Fuzzy set resulting from the "braking" variable obtained via aggregation of rules*

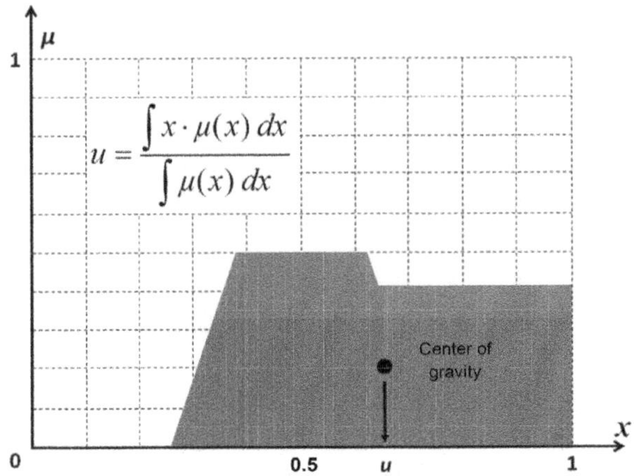

Figure 4.15. *Fuzzy set resulting from the "braking" output variable obtained by aggregating rules*

Defuzzification consists of converting the resulting fuzzy set obtained during the inference phase into a real value (information for an operator, instructions, etc.). The application of the *center of gravity* method is illustrated in Figure 4.15.

4.3. Wind-kinetic energy storage combination on an isolated site with a diesel generator

4.3.1. *Introduction*

The system we will study in this section, combining kinetic energy storage with the production of electrical energy by wind turbines and by a diesel generator on an isolated site (such as an island), illustrates the contribution of fuzzy logic to management adapted to storage, by comparing the proposed supervision solution with a simplified solution based on a simple low-pass filter.

On isolated sites, electrical energy is often produced by generators. However, many of these sites possess exploitable wind potential, and it is therefore of interest to combine some wind turbines with the generators since their operation is more economical due to the no-cost primary source used (wind). However, variations in power due to fluctuating wind-speed combine with variations in power of charges, resulting in a reduced lifespan for the diesel thermal engine. To reduce fuel consumption and the power variations to which the diesel generator is subjected, an energy storage system can be used to absorb these variations. One highly dynamic short-term storage solution is based on a flywheel [LEC 03, CIM 06, ROB 12a, CIM 10].

Figure 4.16 shows the overall operating schema of the energy system examined in this chapter. It is composed of a wind turbine with a fixed speed of 300 kW (which can model several wind turbines of smaller size, for example, six wind turbines of 50 kW each) [ROB 12b]; a 600 kVA diesel generator; a set of charges that are constant for the sake of simplifying our study; and an inertial storage system. The latter is composed of an asynchronous motor of 90 kW with a pair poles connected to the grid via a dual pulse wave modulated (PWM) converter system and a flywheel of 105 kg.m^2. Because the energy stored depends on the square of the flywheel's speed (by the relationship $E = J\omega^2/2$), this speed is maintained at between 3,000 and 6,000 revolutions per minute (RPM).

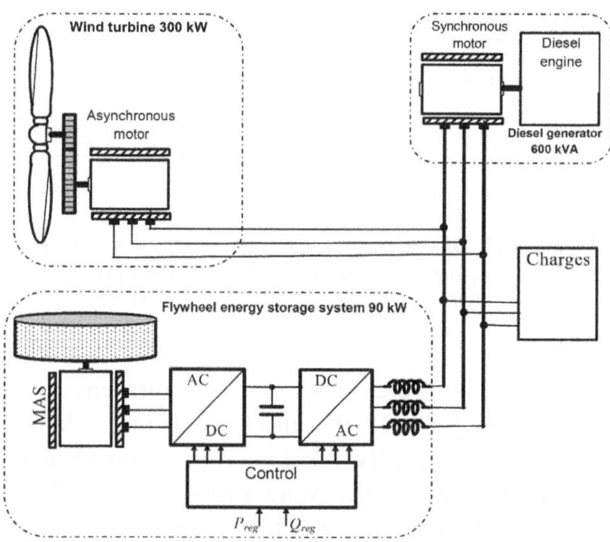

Figure 4.16. *Operating schema of wind-diesel system including flywheel energy storage system*

To control the exchange of power between the grid and the flywheel energy storage system, a fuzzy logic supervisor is proposed, with the aim of reducing the power variations sustained by the diesel generator.

4.3.2. *Energy management strategy*

To reduce the power variations undergone by the diesel generator, the flywheel energy storage system must compensate the variations in the power generated by the wind turbine. If P_{reg} is the power we wish to obtain from the wind-storage system combination (Figure 4.17) and P_{wg} is the power generated by the wind turbine, the reference power that the flywheel energy storage system will have to exchange with the grid is determined as follows:

$$P_{ref} = P_{reg} - P_{wg} \qquad [4.7]$$

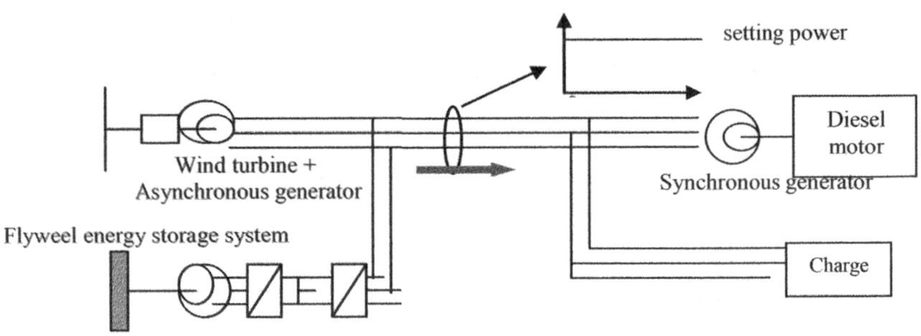

Figure 4.17. *Objective of energy supervision: determination of setting power supplied by combining a wind turbine with flywheel energy storage system*

One problem is that we cannot store or return energy indefinitely due to limited storage size; therefore, we must take this limited capacity into account. For a flywheel energy storage system, the speed of the flywheel determines the state of the storage level (state of charge (SOC)). We must keep to the following reasoning process:

– "if the speed of the flywheel becomes too low then storage is prioritized";

– "if the speed of the flywheel becomes too high then generation is prioritized";

– "if the speed of the flywheel is medium: normal operation".

This, as we can see, is fuzzy logic-type reasoning.

To determine P_{reg}, we use a filtered measurement of the windpower P_{wg}. However, as the speed of the flywheel is limited to between 3,000 and 6,000 RPM, it is necessary to take this speed into account in order to determine P_{reg} so as to avoid saturation of the energy storage system. A fuzzy logic supervisor is thus developed to determine P_{reg}.

4.3.3. *Fuzzy logic supervisor*

Figure 4.18 shows the supervisor input used to determine P_{reg}.

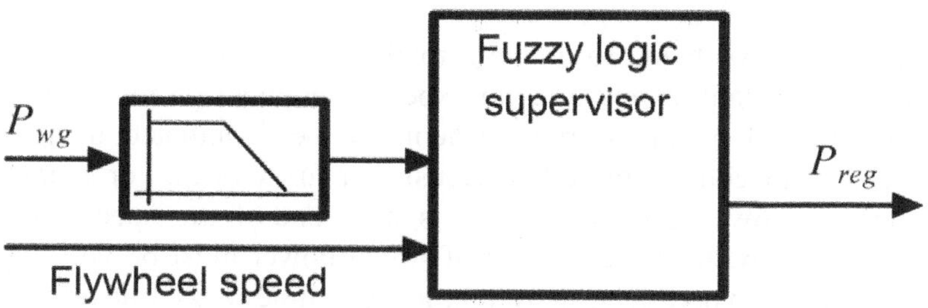

Figure 4.18. *Fuzzy logic supervisor*

4.3.3.1. *Fuzzification*

The membership function of the input variable is represented in Figure 4.19. Three fuzzy states are considered: small (S), medium (M) and big (B).

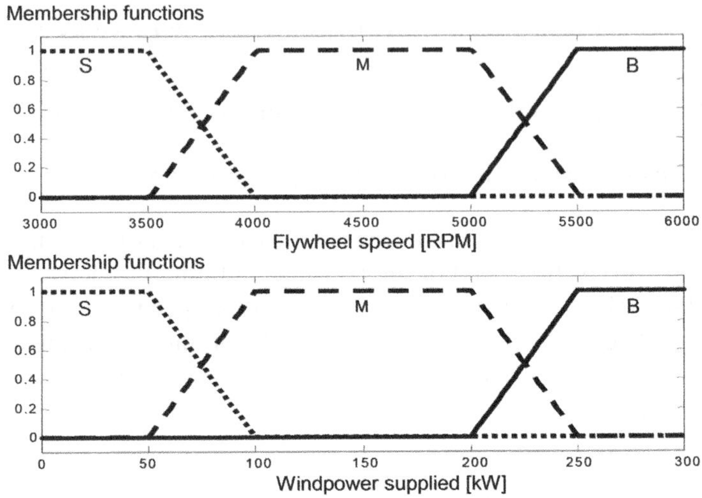

Figure 4.19. *Membership functions of input variables*

4.3.3.2. *Inference*

The objective being to obtain a total generated quantity of power that is as constant as possible, an expert assessment on the operation to be imposed on the storage system can be formulated using the universe of discourse defined previously for fuzzy rules. For example, if the windpower generated (P_{wg}) is high and if the speed of the flywheel is low, then the amount of stored power must be very high. The complete experiment, therefore, leads to nine fuzzy rules that can be expressed as shown in Table 4.2. In order to refine the determination of the adjustment power, seven fuzzy states are considered for the output variable: very small (VS), small (S), small medium (SM), medium (M), big medium (BM), big (B) and very big (VB).

		P_{wg} filtered		
	P_{reg}	Small (S)	Medium (M)	Big (B)
Flywheel Speed	Small (S)	VS	SM	BM
	Medium (M)	S	M	B
	Big (B)	SM	BM	VB

Table 4.2. *Inference table*

The membership function of the output variable is represented in Figure 4.20. Since this variable must be determined with sufficient precision, seven fuzzy sets have been chosen.

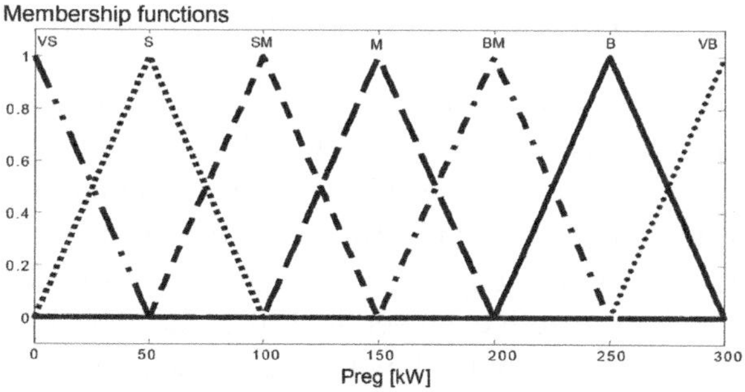

Figure 4.20. *Membership functions of output variable*

4.3.3.3. *Defuzzification*

Figure 4.21 represents the evolution of the output variable, the adjustment power, depending on the input variables of filtered windpower and flywheel speed; these variables have been normalized in per unit in relation to the maximum value each of them can take. This evolution is nonlinear. Note that for each pair of input variables there is a single corresponding value for the output variable; therefore, fuzzy logic is determinist.

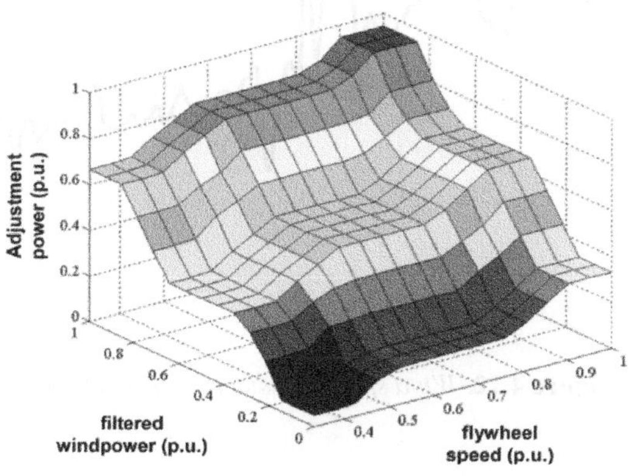

Figure 4.21. *Evolution of output variable versus input variables. For a color version of the figure, see www.iste.co.uk/robyns/powergrids.zip*

Inference was accomplished by means of the PROD/SUM method and defuzzification using the center of gravity method.

4.3.4. *Results of simulation with fuzzy supervisor [ROB 12a]*

To simulate a wind turbine, measurements were taken on a windmill with a constant speed of 300 kW, installed in the north of France. Figure 4.22 shows the wind speed measured in the case of medium wind speed (around 10 m/s) followed by low speed (around 6.3 m/s). In the study, the active power consumed by the load is 300 kW and its reactive power is 120 kVar. The initial rotation speed of the flywheel coupled with the asynchronous generator is 3,500 RPM.

Figure 4.23 shows the active power generated by the diesel generator with and without a flywheel energy storage system as a solid line and a dotted line, respectively. This figure confirms that the proposed storage system enables a significant reduction in the diesel generator's power variations. During sudden variations in wind speed, significant power variations persist.

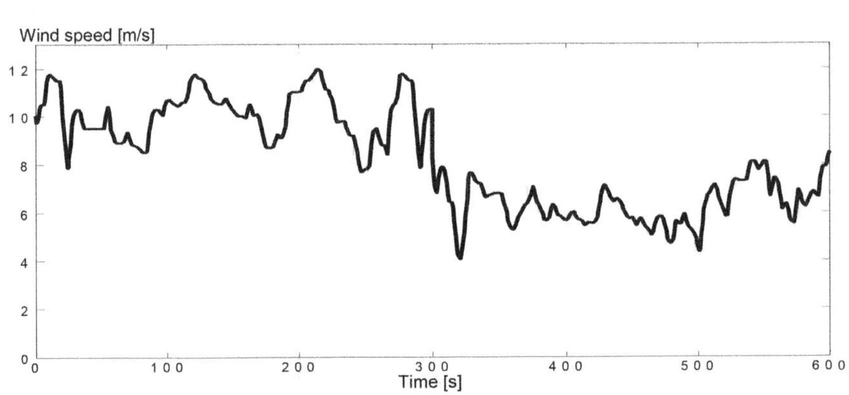

Figure 4.22. *Wind speed considered in simulations*

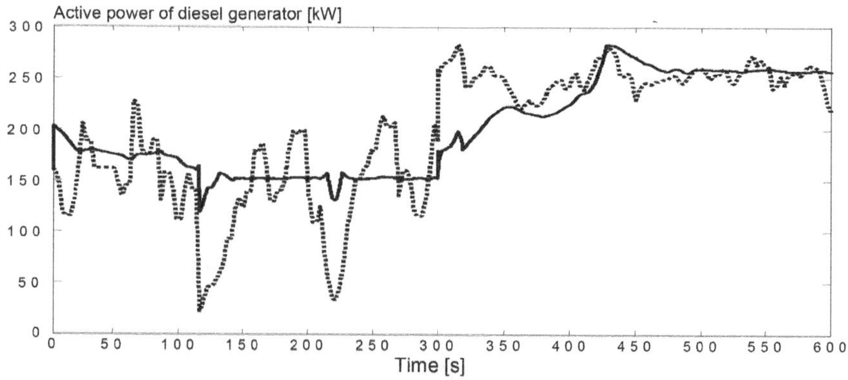

Figure 4.23. *Active power generated by diesel generator with and without inertial storage system in solid and dotted lines, respectively*

Figure 4.24 shows that these variations appear when the desired power P_{reg} (shown as a dotted line) cannot be maintained by the system considered, for which the power generated by the wind turbine-storage system pairing is represented as a solid line. This is due to the power of the asynchronous motor coupled with the flywheel, which is limited to 90 kW.

The rotation speed of the flywheel is shown in Figure 4.25. Saturation speed does not appear, as this speed has been taken into account in the determination of the reference power (Table 4.2).

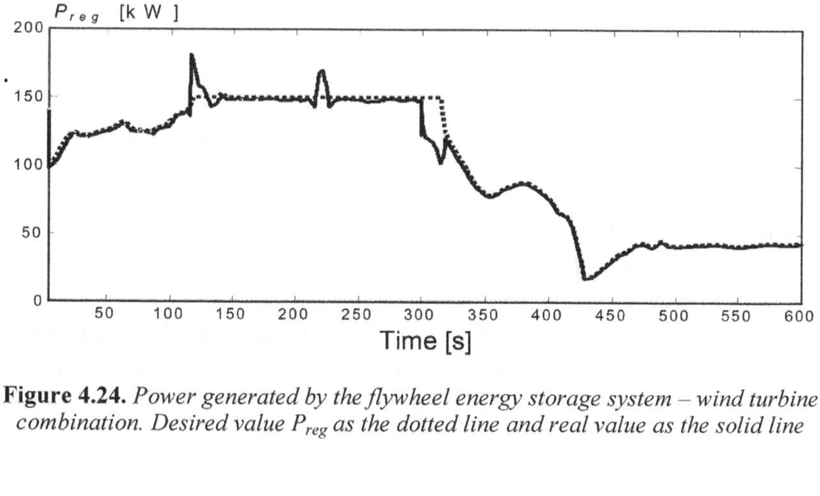

Figure 4.24. *Power generated by the flywheel energy storage system – wind turbine combination. Desired value P_{reg} as the dotted line and real value as the solid line*

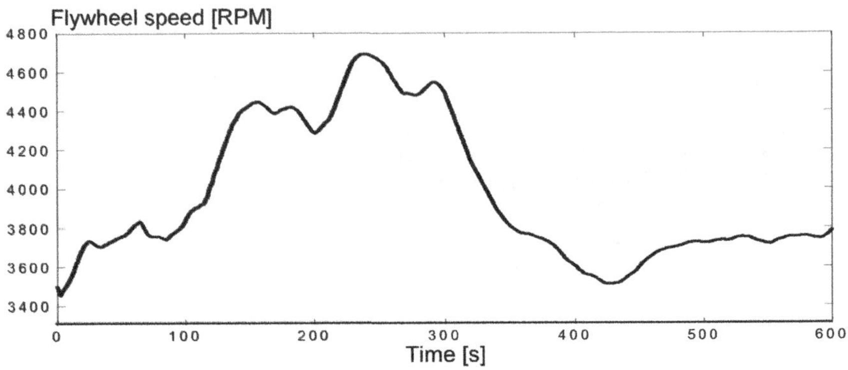

Figure 4.25. *Flywheel speed*

4.3.5. *Results of simulation with simple filtering*

Since the objective of inertial storage system supervision is to smooth the windpower generated, we can reduce the supervisor from Figure 4.18 to a simple low-pass filter. This gives us the supervisor represented in Figure 4.26. The simulations presented in this section have been conducted assuming the same time constant for the filter as in the previous section, that is 30 s; the same wind profile (Figure 4.22); and the same initial speed for the level of storage (flywheel speed of 3,500 RPM).

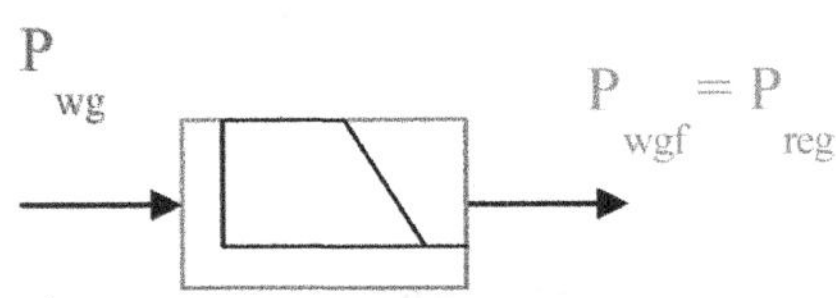

Figure 4.26. *Supervision carried out by means of a low-pass filter*

Figure 4.27 shows the active power generated by the diesel generator with and without an inertial storage system as a solid line and a dotted line, respectively. Compared to Figure 4.23, the power generated in the presence of storage system appears less smoothed, and the areas where this power is virtually constant are less numerous. Significant power variations persist.

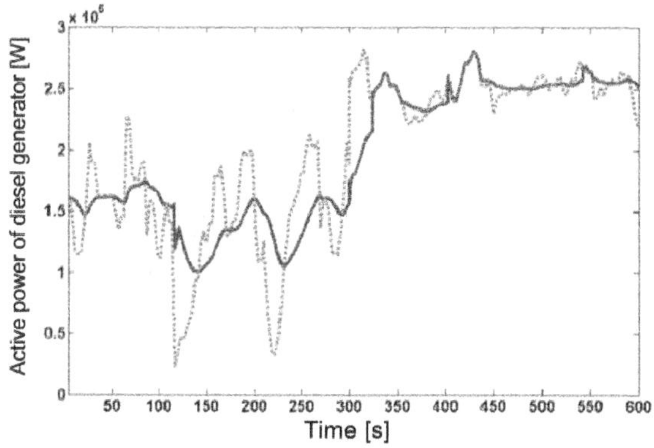

Figure 4.27. *Active power generated by diesel generator with and without a flywheel energy storage system in the solid and dotted lines, respectively. Supervision by filtering*

Figure 4.28 shows the difference between the desired power P_{reg} (shown as a dotted line) and the power actually generated by the wind turbine–storage system (shown as a solid line) when the wind speed is low. Unlike the result shown in Figure 4.24, variations in the desired setting power are slower, and the limit power of the asynchronous motor coupled with the flywheel is not questioned.

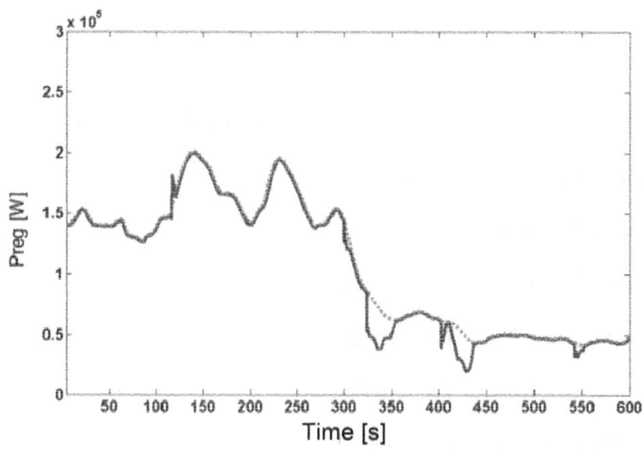

Figure 4.28. *Power generated by the flywheel energy storage system–wind turbine combination. Desired value P_{reg} is shown as a dotted line and real value as a solid line. Supervision by filtering*

The rotation speed of the flywheel is shown in Figure 4.29. This speed saturates at its low value when the wind speed is low. The storage system has in fact reached its lowest acceptable energy level. In order to avoid this saturation, it would be necessary to increase the energetic capacity of the storage system, or to take into account the level of storage as proposed in the previous section, inducing an evolution of the unsaturated storage level like the one shown in Figure 4.25.

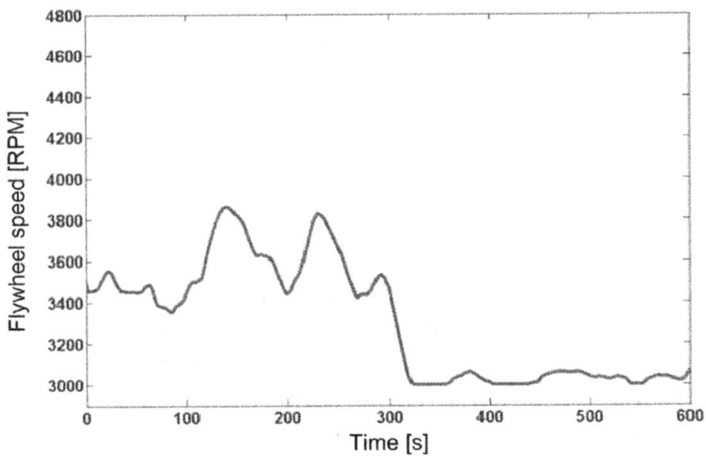

Figure 4.29. *Flywheel speed. Supervision by filtering*

Parameters of asynchronous motor coupled with flywheel are:

– nominal power: 90 kW;

– number of poles: 2;

– stator resistance: 15.8 mΩ;

– rotor resistance: 14.5 mΩ;

– mutual inductance: 18.9 mH;

– stator inductance: 19.2 mH.

4.4. Conclusion

Fuzzy logic has been used to build a multiobjective supervision strategy, with two main objectives: to smooth windpower while managing energy storage in order to avoid saturation and to ensure its maximum availability to absorb or generate energy. This approach enables the best use of storage capacity by reducing the size of that system.

The supervisor developed is nonlinear, but ensures ongoing gentle transitions between system states. It is determinist and of limited complexity. The number of fuzzy rules depends on the number of inputs and the number of fuzzy sets chosen for each input; in the example developed in section 4.3.3, this number is 3.3, i.e. 9 rules.

Chapter 5 introduces a methodology used systematically to design a fuzzy supervisor while limiting the empiricism of the approach and controlling the complexity of the algorithm obtained.

The fuzzy reasoning developed in this chapter to manage a flywheel energy storage system can, of course, be transposed to other storage technologies.

4.5. Bibliography

[BOR 98] BORNE P., RORINOER J., DIEULOT J-Y. *et al.*, *Introduction à la commande floue*, Technip, 1998.

[BOU 98] BOUCHON-MEUNIER B., FOULLOY L., RAMDANI M., *Logique floue. Exercices corrigés et exemples d'applications*, Capaduès, 1998.

[BUH 94] BUHLER H., *Réglage par logique floue,* Presses Polytechniques et Universitaires Romandes, 1994.

[CIM 06] CIMUCA G., SAUDEMONT C., ROBYNS B. *et al.*, "Control and performance evaluation of a flywheel energy storage system associated to a variable speed wind generator", *IEEE Transactions on Industrial Electronics*, vol. 53, no. 4, pp. 1074–1085, August 2006.

[CIM 10] CIMUCA G., BREBAN S., RADULESCU M. *et al.*, "Design and control strategies of an induction machine-based flywheel energy-storage system associated to a variable-speed wind generator", *IEEE Transactions on Energy Conversion*, vol. 25, no. 2, pp. 526–534, June 2010.

[LEC 03] LECLERCQ L., ROBYNS B., GRAVE J.M., "Control based on fuzzy logic of a flywheel energy storage system associated with wind and diesel generators", *Mathematics and Computers in Simulation*, Elsevier, vol. 63, pp. 271–280, 2003.

[ROB 12a] ROBYNS B., BRUNO F., DEGOBERT P. *et al.*, *Vector Control of Induction Machines. Desensitisation and Optimisation Through Fuzzy Logic*, Springer Verlag, 2012.

[ROB 12b] ROBYNS B., DAVIGNY A., BRUNO F., *et al. Electric Power Generation from Renewable Sources*, ISTE, London and John Wiley & Sons, New York, 2012.

5

Supervisor Construction Methodology for a Windpower Source Combined with Storage

5.1. Introduction

The increased presence of decentralized production in electric grids is due in part to the technological evolution of these sources. New generations will have to be capable of participating in ancillary services as explained in Chapter 3: adjustment of voltage, frequency and reactive power; ability to start independently and operate in isolated mode; etc. [ROB 12].

Due to the highly variable and random nature of the wind source, as well as errors in predicting the associated production of electricity, a wind turbine alone can participate only partially in ancillary services, provided that adapted supervision is implemented [MOK 09a, MOK 09b]. Its coupling with an energy reserve, as well as the development of adequate supervision, is a necessary solution [DAV 06]. The characteristics of inertial storage are well suited to this application in terms of dynamics, lifespan, efficiency, etc.

The developments introduced in this chapter have to do with the combination of an inertial energy storage system with a variable-speed wind generator, resulting in an ensemble capable of providing ancillary services or operating in isolated mode [CIM 05, DAV 06].

We will show that classic solutions exist that enable the proposed production system to attain an objective and then we will examine the development of energy management strategies incorporating these tools as well as fuzzy logic. The first seven stages of the methodology introduced in Chapter 1 will be applied in the development of a multiobjective energy supervisor.

An experimental application will be used to discuss the real-time implantation of this type of supervisor. Experimental tests will be used to compare different variants of supervisors and to validate the theoretical developments previously presented.

5.2. Energetic system studied

Figure 5.1 represents the operating schema of a variable-speed wind turbine constructed on the basis of a permanent magnet synchronous generator (PMSG). The flywheel energy storage system (FESS), composed of a flywheel combined with a variable-speed cage asynchronous generator, is coupled with the direct current (DC) bus [LEC 03, CAR 01].

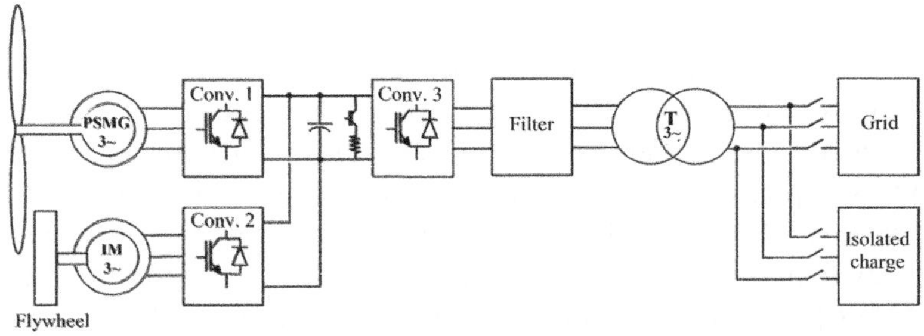

Figure 5.1. *Variable-speed wind turbine combined with an inertial storage system*

The interconnection of the various elements (wind generator, electrical generator combined with a flywheel and connection to electric grid) is done by means of alternative current (AC)-DC electronic power converters, which are completely controllable and connected to a single DC bus. This system can function when connected to an electric grid [CIM 05] or in an isolated mode [DAV 06].

5.3. Supervisor development methodology

A graphic method is used to design the supervisor. This methodology requires no mathematical models, as it is based on the expertise of the system represented by fuzzy rules. Inputs can be random and supervision can target multiple objectives simultaneously. Transitions between operating modes are progressive since they are determined by fuzzy variables. Finally, this methodology enables storage management via a convergence toward a state of charge and control of complexity with a view to real-time processing. The design methodology of the supervision system applied in this case involves seven stages:

1) determination of system specifications; the characteristics and objectives of the system must be clearly defined, as well as its limitations and means of action;

2) structure of the supervisor: the necessary inputs and outputs are determined;

3) determination of "functional graphs", a graphic representation of the operating modes is proposed, based on knowledge of the system;

4) determination of the membership functions of the fuzzy supervisor;

5) determination of "operational graphs"; a graphic representation of the fuzzy operating modes is proposed;

6) fuzzy rules, characteristics of the fuzzy supervisor, are extracted from the operational graphs;

7) definition of indicators used to assess the achievement of objectives.

An experimental phase is used to validate the supervisor developed and to refine the values of the parameters used in the supervisor.

In this study, storage dimensions are assumed to be predetermined (an industrial system offered in a builder's catalog is used).

5.4. Specifications

For the energetic system shown in Figure 5.1, we must identify the objectives and limitations, as well as the means of action.

5.4.1. *Objectives*

We will consider the case of the operation of a windpower production system connected to the electric grid.

The first objective is to send smoothed power back to the grid, which is hypothetically able to receive the power sent by the wind turbine.

The second objective is to ensure the availability of storage; that is to ensure that it does not reach its high or low limits (for the latter case, we nevertheless assume minimal wind speed). Thus, we are permanently managing the energy level contained in the storage system without expecting a high or low limit to be reached; this will prevent any saturation of the storage system.

Finally, we must consider the question of maintaining the voltage of the DC bus.

A wind turbine connected to the grid via the grid connection converter (Figure 5.1, labeled as *Conv3*) may contribute to this maintenance. However, if we wish this wind turbine to participate in ancillary services, the converter must be freed from performing this task. Maintenance of the DC bus is assigned to the FESS.

5.4.2. *Limitations*

We have hypothesized that the electric grid is capable of receiving the power injected; therefore, there is no limitation to be assumed at this level. On the other hand, power smoothing is dependent on storage capacity.

Next, to ensure the correct functioning of the system, and more particularly of the electronic power converters, the DC bus voltage must be maintained at a fixed value in permanent operation.

Finally, the inertial storage system has limits in terms of high and low energy. The kinetic energy E_c stored in a flywheel results in the rotation speed Ω, the inertia of which is J, written as:

$$E_c = \frac{1}{2} \cdot J \cdot \Omega^2 \qquad [5.1]$$

The upper limit is imposed by the drive machine, chosen so that this limit speed will be slower than the maximum mechanical speed acceptable for the flywheel.

The lower limit is theoretically chosen in order to maintain the asynchronous generator in its operating zone at constant power, thus enabling an adequate efficiency to be maintained.

If we use Ω_{min} and Ω_{max}, respectively, to denote the low and high drive speeds of the flywheel, the variation of the energy contained in the storage device, called ΔE, is written as:

$$\Delta E = \frac{1}{2} \cdot J \cdot \Omega^2 \max \cdot \left(1 - \frac{\Omega \min}{\Omega \max}\right) \qquad [5.2]$$

To increase the quantity of energy exchanged between the storage device and the rest of the system, we can reduce the value chosen for Ω_{min}.

5.4.3. *Means of action*

In our case study, the PMSG is controlled so as to extract the maximum amount of power; therefore, we will not attempt to reduce the power generated by the wind turbine. Likewise, blade angle calibration is not planned in normal operating mode [ROB 12].

Thus, two means of action are identified: the base power to be injected into the grid, which we will write as P_{reg}, and the base power designated for the storage system P_{FESS_ref}.

Table 5.1 summarizes the specifications of the system being studied.

Objectives	Constraints	Means of action
Smoothing of power sent to grid Availability of storage Maintenance of DC bus voltage	Limited storage capacity Lower and upper storage limits DC voltage setpoint value	Base power injected into the grid: P_{reg} Base storage power: P_{FESS_ref}

Table 5.1. *Summary of supervisor specifications*

5.5. Supervisor structure

At this point, we will determine the supervisor's input and output values.

5.5.1. *Input values*

Supervisor input depends on the objectives set. Each objective corresponds to at least one input.

The first objective is to smooth the power generated by the wind turbine in order to inject it into the electric grid. Thus, the first input is the windpower P_{wg}.

The second objective is to maintain storage availability between its upper and lower limits. The supervisor must be informed of the energy contained in the flywheel. According to equation [5.1], the measurement of the rotation speed, Ω, is necessary.

Finally, the voltage of the DC bus must be maintained; the measurement of this voltage, written as V_{dc}, is also a supervisor input value.

5.5.2. *Output values*

Outputs are the references sent to the means of action; or P_{reg} and P_{FESS_ref}, respectively.

The structure of the supervisor is shown in Figure 5.2.

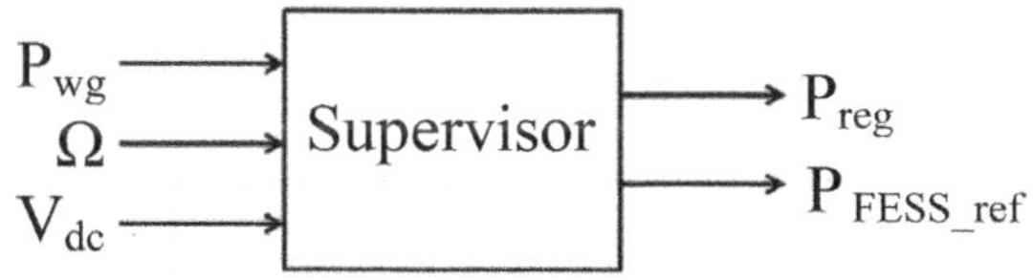

Figure 5.2. *Supervisor of a variable-speed wind turbine combined with a flywheel storage system*

At this point in the development of the supervisor, it is useful to identify the various tools to be used.

5.5.3. *Supervisor development tools*

The smoothing of the power to be sent to the grid requires the presence of a filter, and thus of classic frequential processing, as proposed in [JAA 09]. The principle is to send power P_{reg} to the grid, corresponding to the low-frequency component originating from the windpower, P_{wg_LF} [5.3].

$$P_{reg} = P_{wg\_LF} \quad [5.3]$$

The high-frequency component of the windpower, P_{wg_HF}, is thus designated for storage, which will have to compensate for it by storing or generating energy.

Finally, since the voltage V_{dc} must be enslaved to a fixed setpoint value, a simple and classic proportional integral (PI) corrector is used.

Insofar as this voltage maintenance is assigned to the storage system, the power ΔP necessary for this maintenance constitutes an element of the base power of the storage system.

This base power can, therefore, be written as [5.4]:

$$P_{FESS\_ref} = P_{wg\_HF} + \Delta P \qquad [5.4]$$

At this stage of development, the supervisor is as shown in Figure 5.3. It uses basic tools (first-order low-pass filter and PI corrector) and can ensure that the objectives of smoothing the power sent back to the grid and maintaining the voltage of the DC bus will be fulfilled.

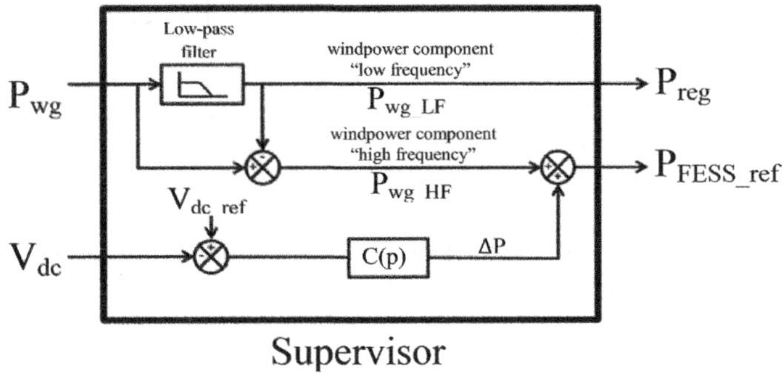

Figure 5.3. *Supervisor fulfilling two objectives: smoothing of the power sent back to the grid and maintenance of DC bus voltage*

However, in this state, it cannot ensure the availability of storage. This case corresponds to the case illustrated in Figure 4.29, in which we can see a low-value saturation of the flywheel speed.

To fulfill this final objective, the measurement of the flywheel rotation speed Ω must be taken into account by the supervisor and contribute to the determination of the power to be sent to the grid. Fuzzy logic is used here due to the highly variable aspect of Ω and P_{wg_LF}, which are thus the two inputs of the fuzzy part of the supervisor (Figure 5.4), which we will henceforth call the fuzzy supervisor.

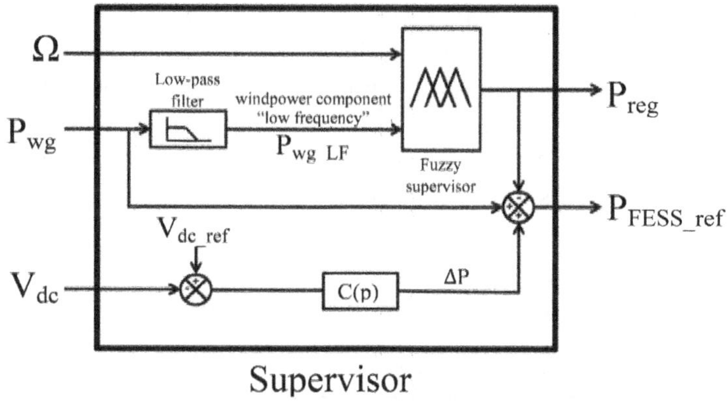

Figure 5.4. *Supervisor fulfilling all three objectives: smoothing of power sent to the grid; control of DC bus voltage; and availability of storage*

The base power intended for storage, P_{FESS_ref}, is expressed according to relationship [5.5], while the base power to be sent to the grid is determined by the fuzzy supervisor.

$$P_{FESS\_ref} = P_{wg} + \Delta P - P_{reg} \tag{5.5}$$

Table 5.2 shows a summary of the specifications, completed by the tools used in developing the supervisor.

Objectives	Limitations	Means of action	Tools
Smoothing of power sent to grid	Limited storage capacity	Base power injected into grid: P_{reg}	Low-pass filter
Availability of storage	Lower and upper storage limits		Fuzzy logic
Maintenance of DC bus voltage	DC voltage setpoint value	Base storage power: P_{FESS_ref}	PI corrector

Table 5.2. *Summary of supervisor specifications including tools*

The elements of the supervisor shown in Figure 5.4 that must be precisely defined are the parameters of the PI corrector and the frequency of the low-pass filter and the fuzzy supervisor.

The parameters of the PI corrector, applied to the measurement of the DC bus voltage, are determined using classic methods from automatics. Figure 5.5 shows this type of structure, in which the upper part represents the system to be controlled (first order, where K_{sys} and τ_{sys} represent the system gain and time constant, respectively), and the lower part represents the PI corrector.

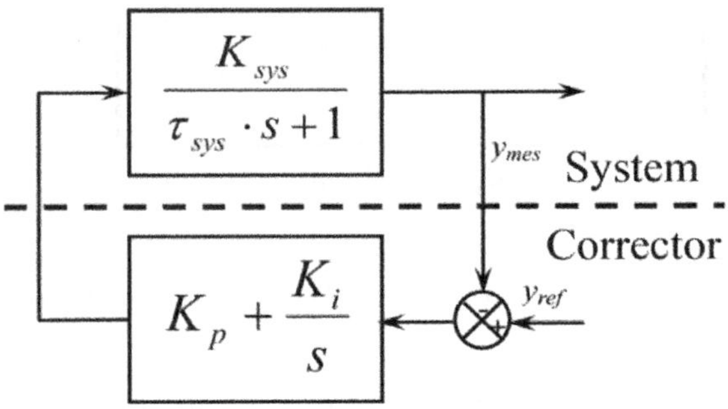

Figure 5.5. *Proportional integral corrector*

In this study, the controlled system is modeled by an equivalent capacity C, and a leak resistance R_c, with the controlled variable being the DC bus voltage, V_{dc} (Figure 5.6).

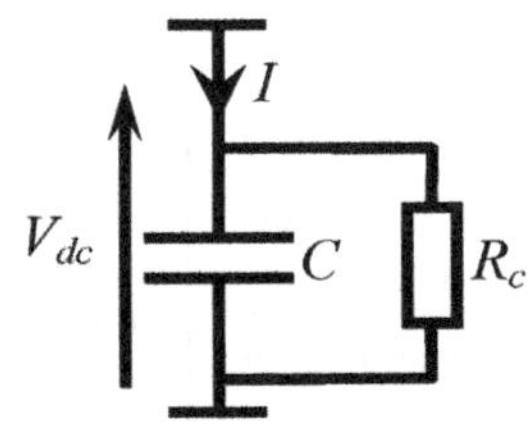

Figure 5.6. *System to be controlled*

The transfer function of this system is as follows:

$$\frac{V_{dc}}{I} = \frac{R_c}{C \cdot R_c \cdot s + 1} \qquad [5.6]$$

The time constant τ of the filter is usually determined as a function of the storage capacity [CIM 05, JAA 09]. In this case, we use relationship [5.7], where J is the flywheel inertia, P_n is the nominal power that can be sent to or generated by storage and Ω_{max} and Ω_{min} are the maximum and minimal speeds, respectively, of the flywheel.

$$\tau = \frac{J \cdot \left(\Omega_{max}^2 - \Omega_{min}^2\right)}{2 \cdot P_n} \quad [5.7]$$

The rest of the chapter is devoted to the development of the fuzzy supervisor, and the first point addressed will be the identification of the various operating states of the system in order to create the functional graph.

5.6. Identification of various operating states: functional graph

The fuzzy supervisor has two inputs (flywheel rotation speed, storage state of charge and the low-frequency component of the wind-turbine power) and one output, which is the base power to be sent to the grid (Figure 5.7). Note that the structure of this fuzzy supervisor is identical to the structure of the supervisor developed in Chapter 4 and shown in Figure 4.18.

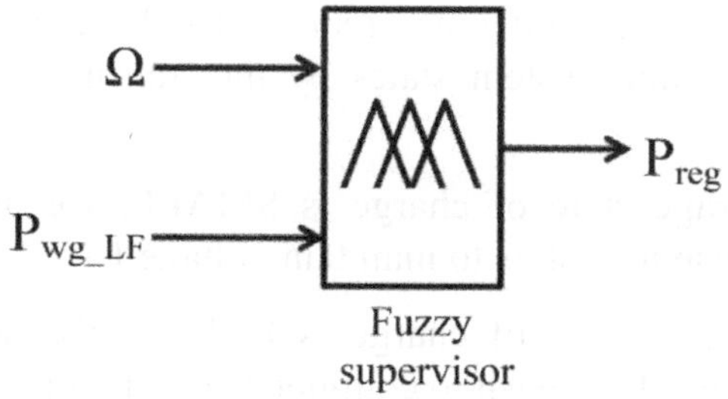

Figure 5.7. *Fuzzy supervisor inputs and output*

At this point, we must determine the power to send to the grid (production system output), depending on the state of charge of the storage system and the windpower supplied.

The strategy of the fuzzy supervisor can be defined graphically; this graphic definition has the following advantages:

– a literal expression of the objectives and subobjectives to be fulfilled, constraints and means of actions permitting:

- the direct establishment of the fuzzy laws applicable to each operating mode and thus, the limitation of the supervisor's complexity;

- the facilitation of exchanges with other disciplinary fields such as the economy, which plays a significant role in energy choices.

– a transition between modes determined by the state of certain parts of the system. These states can be described by the fuzzy variables that are the supervisor's inputs, thus enabling smooth transitions between modes of operation and the system's ability to function in multiple modes simultaneously;

– insofar as fuzzy logic integrates Boolean logic, it can be used to revisit classic approaches such as Petri or Grafcet grids.

5.6.1. *Graph of level N1*

The first supervisor level N_1 has the objective of sending windpower to the grid. Operating states are indicated on the graph in Figure 5.8. Working modes are represented by the rectangles with rounded corners and system states by the transitions between these modes:

– if the storage state of charge is SMALL, we must attempt to restore it, because its task is to maintain voltage V_{dc};

– if the storage state of charge is LARGE, we must attempt to preserve a margin if we wish to continue smoothing power;

– if the storage state of charge is MEDIUM, no specific action is advised.

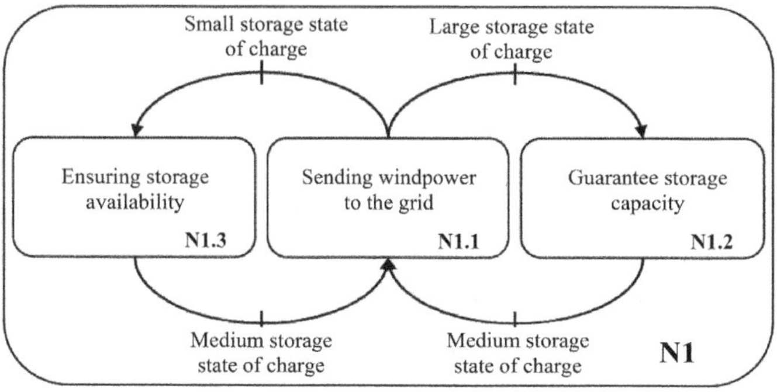

Figure 5.8. *Functional graph of level 1*

5.6.2. *Graph of level N1.1*

Level N1.1 (Figure 5.9) corresponds to system operation when there is a MEDIUM state of charge.

No specific action in terms of charge or discharge is advised with regard to storage.

The tendency is, therefore, to send power to the grid that is equivalent to the windpower supplied by the generator.

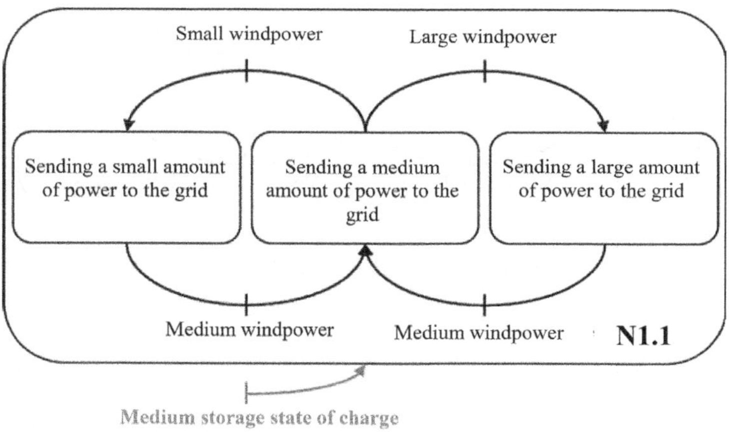

Figure 5.9. *Functional graph of level 1.1*

5.6.3. *Graph of level N1.2*

Level N1.2 (Figure 5.10) corresponds to system operation when there is a LARGE storage state of charge.

This means that the tendency is to send an amount of power to the grid that is larger than the windpower in order to draw on the storage and thus, contribute to ensuring its availability:

– if the windpower is SMALL, the system is requested to send slightly higher power (MEDIUM SMALL) to the grid;

– if the windpower is LARGE, the system is requested to send VERY HIGH power to the grid in order to avoid "high" saturation of storage;

– if the windpower is MEDIUM, the system is requested to send slightly higher power (MEDIUM LARGE) to the grid.

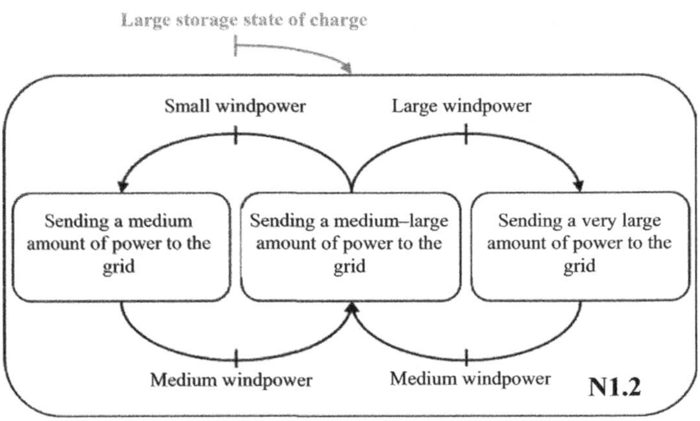

Figure 5.10. *Functional graph of level 1.2*

5.6.4. *Graph of level N1.3*

Level N1.3 (Figure 5.11) corresponds to system operation when there is a SMALL storage state of charge.

The trend is thus to send power to the grid that is lower than the windpower in order to partially top up the storage:

– if the windpower is SMALL, the amount of power sent to the grid is VERY SMALL;

– if the windpower is LARGE, the amount of power sent to the grid is MEDIUM LARGE;

– if the windpower is MEDIUM, the amount of power sent to the grid is MEDIUM SMALL.

The behavior of the production system according to the charge state of the storage device and to windpower having been determined, we will now proceed to the determination of the membership functions linked to the input and output variables of the fuzzy supervisor.

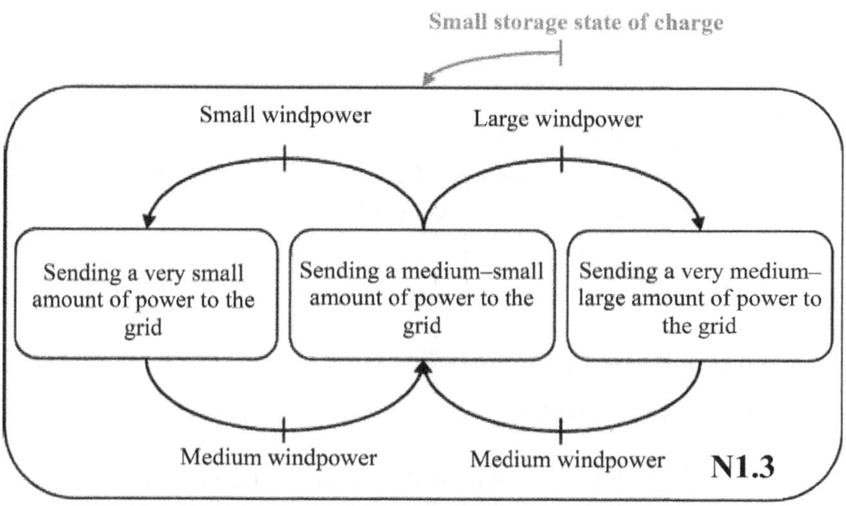

Figure 5.11. *Functional graph of level 1.3*

5.7. Membership functions

With regard to Figure 5.7, the variables to process in this part are the two input variables, P_{wg_LF} and Ω, and the output variable P_{reg}.

Input values are normalized [5.8] and appear on the graphs in *per unit* (p.u.):

$$\Omega[\text{p.u.}] = \frac{\Omega}{\Omega_{max}}$$

$$P_{wg\_LF}[\text{p.u.}] = \frac{P_{wg\_LF}}{P_{wg\_LF_{max}}} \qquad [5.8]$$

It is convenient to represent the *linguistic variable* corresponding to each of these values by a trio of information containing the name of the variable, its universe of discourse and the set of possible *fuzzy characterizations* the fuzzy variable may take:

$$\{\textbf{\textit{Name of linguistic variable, universe of discourse, set of characterizations}}\}$$

The terms used to characterize the variable are called *linguistic labels*.

In this case, the minimum drive speed of the flywheel is 800 rpm, while the maximum speed is 3,000 rpm; this allows a maximal variation of the energy exchanged between the storage and the rest of the system of around 0.93 E_{max}, according to relationship [5.2].

The universe of discourse in *p.u.* of the variable Ω is, therefore, included between 8/30 and 1 (Figure 5.12).

The variables are defined as follows:

– for the input variables:

$$\left\{\Omega, \left[\frac{8}{30}, 1\right], \text{"SMALL"}, \text{"MEDIUM"}, \text{"BIG"}\right\}, \text{ (Figure 5.12)}$$

$$\left\{P_{wg\_LF}, [0,1], \text{"SMALL"}, \text{"MEDIUM"}, \text{"BIG"}\right\}, \text{ (Figure 5.13);}$$

– for the output variable of the fuzzy supervisor:

$$\{P_{reg}, [0,1], \text{"VERY SMALL"}, \text{"SMALL"}, \text{"MEDIUM SMALL"}, \text{"MEDIUM"}, \text{"MEDIUM BIG"}, \text{"BIG"}, \text{"VERY BIG"}\}, \text{ (Figure 5.14).}$$

For the sake of clarity of the figures, the linguistic labels are assigned to them using their initials as shown in Table 5.3.

Linguistic labels	VERY SMALL	SMALL	MEDIUM–SMALL	MEDIUM	MEDIUM–BIG	BIG	VERY BIG
Simplified linguistic labels	VS	S	MS	M	MB	B	VB

Table 5.3. *Linguistic labels and simplifications*

Three fuzzy sets are chosen for each input variable. As is reiterated in the literature [BOR 98, BUH 94], there is no set rule for choosing this number. Three, five or seven fuzzy sets are generally found; so a compromise must be made between the precision of the description of the variable (which leads to a large number of fuzzy sets) and an increase in the number of fuzzy rules.

The three membership functions of the flywheel drive speed are chosen by experiment so as to be trapezoidal and symmetric in relation to the universe of discourse.

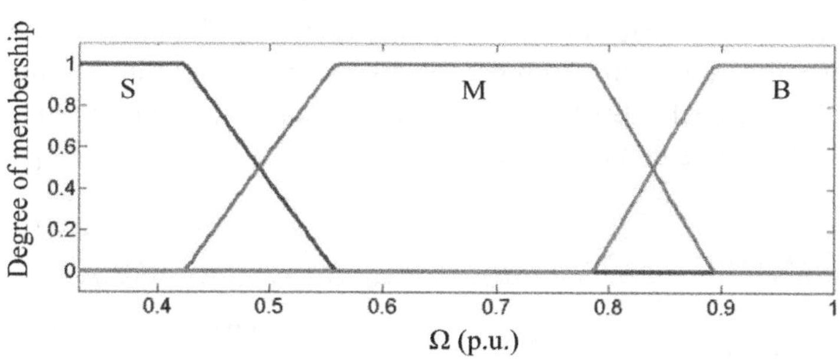

Figure 5.12. *Membership functions of flywheel speed Ω. For a color version of the figure, see www.iste.co.uk/robyns/powergrids.zip*

Because filtered windpower can vary from zero to nominal power, the universe of discourse of the variable P_{wg_LF} varies between 0 and 1. The appearances of the membership functions are chosen by experiment.

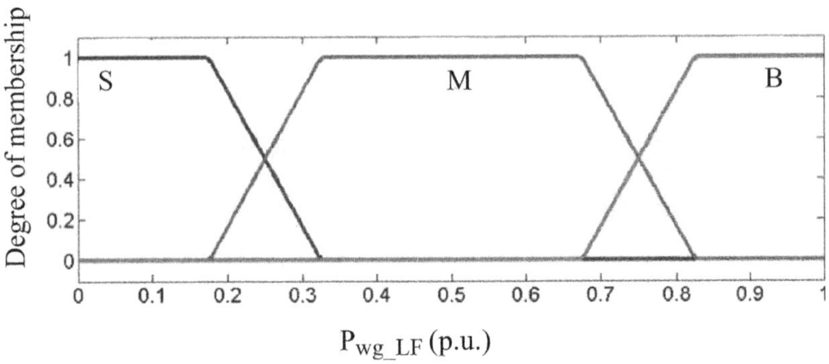

Figure 5.13. *Membership functions of filtered windpower P_{wg_LF}. For a color version of the figure, see www.iste.co.uk/robyns/powergrids.zip*

The fuzzification of the two variables Ω and P_{wg_LF} is based on the use of three membership functions, compromising between precision and limitation of the number of fuzzy rules, determined by the product of the numbers of fuzzy sets of each input variable multiplied together (or, in this case, $3 \times 3 = 9$ rules).

The same operation on the output variable P_{reg} uses seven membership functions in order to soften the changes of state of this variable. This large number of membership functions does not increase the maximum number of fuzzy rules. Insofar as the input variables are each fuzzified using three functions, a maximum of nine fuzzy sets can be considered for P_{reg}.

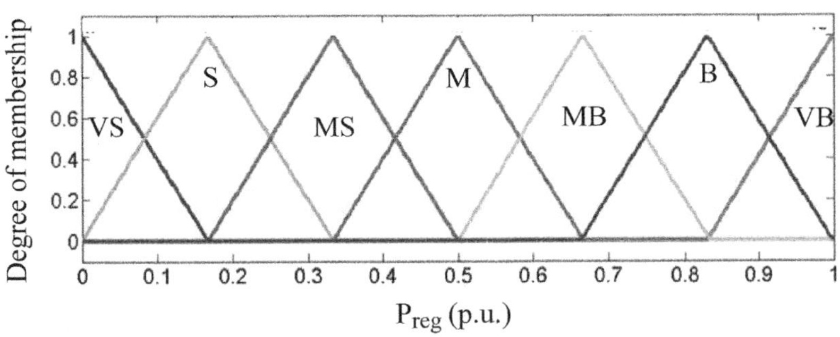

Figure 5.14. *Membership functions of power sent to the grid P_{reg}. For a color version of the figure, see www.iste.co.uk/robyns/powergrids.zip*

The functional graph and membership functions of the input and output variables of the fuzzy supervisor having been defined, we can now develop the operational graph.

5.8. Operational graph

To determine the fuzzy laws, it is necessary to translate the functional graphs into operational graphs that include the membership functions previously defined. Transitions between operational modes will be described by the membership functions of the input values, and the actions of the operational modes by the membership functions of the output values.

The principal operational graph and operational subgraphs are shown in Figures 5.15–5.18.

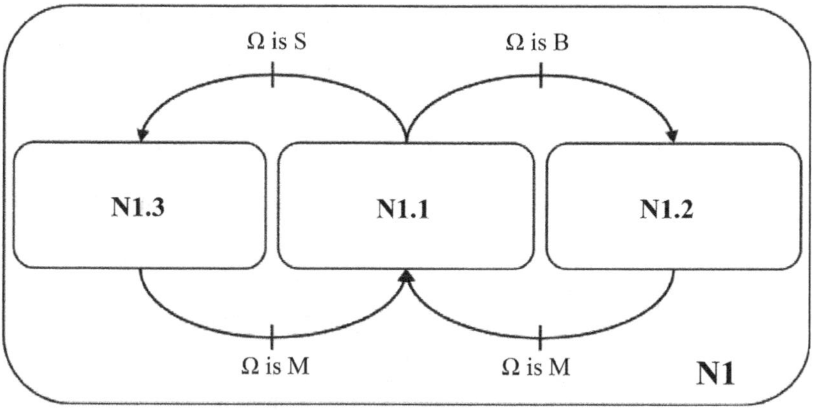

Figure 5.15. *Operational graph of level 1*

Creating them involves using variable names and simplified linguistic labels for both states and transitions.

Thus, for example, "low windpower" becomes "P_{wg_LF} is S", and "sending a Medium Big amount of power to the grid" becomes "P_{reg} is MB", etc.

With regard to the operation of the algorithm, unlike what is observed in a tool such as Grafcet, passage from one action to another happens progressively.

5.8.1. *Graph of level N1*

In the graph shown in Figure 5.15, we see the three levels N1.1–N1.3 activated, with progressive transitions, depending on the state of charge of the storage device.

5.8.2. *Graph of level N1.1*

This level (Figure 5.16) is activated when the state of charge of the storage device is MEDIUM ("Ω is M"). The graph shows the correspondence between the operations considered in the state graphs (Figure 5.9) and the linguistic values assigned to the variables:

– "sending a MEDIUM amount of power to the grid" versus "P_{reg} is M";

– "sending a BIG amount of power to the grid" versus "P_{reg} is B";

– "sending a SMALL amount of power to the grid" versus "P_{reg} is S".

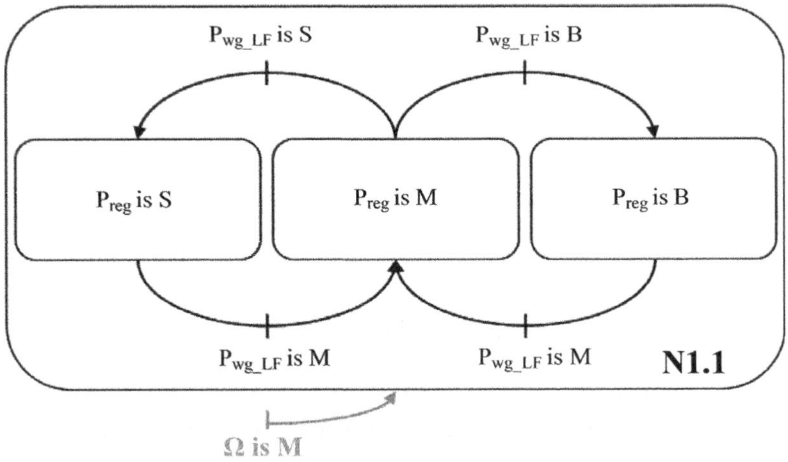

Figure 5.16. *Operational graph of level 1.1*

5.8.3. *Graph of level N1.2*

This level (Figure 5.17) is activated when the state of charge of the storage device is BIG ("Ω is B"). The graph shows the correspondence between the operations considered in the state graphs (Figure 5.10) and the linguistic values assigned to the variables:

– "sending a MEDIUM BIG amount of power to the grid" versus "P_{reg} is MB";

– "sending a VERY BIG amount of power to the grid" versus "P_{reg} is VB";

– "sending a MEDIUM SMALL amount of power to the grid" versus "P_{reg} is MS".

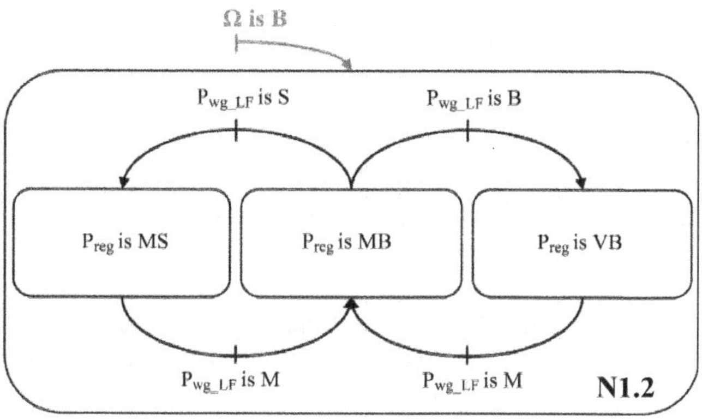

Figure 5.17. *Operational graph of level 1.2*

5.8.4. *Graph of level N1.3*

This level (Figure 5.18) is activated when the state of charge of the storage device is SMALL ("Ω is S"). In the graph, we can see the correspondence between the operations considered in the graphs of state (Figure 5.11) and the linguistic values assigned to the variables:

– "sending a MEDIUM SMALL amount of power to the grid" versus "P_{reg} is MS";

– "sending a MEDIUM BIG amount of power to the grid" versus "P_{reg} is MB";

– "sending a VERY SMALL amount of power to the grid" versus "P_{reg} is VS".

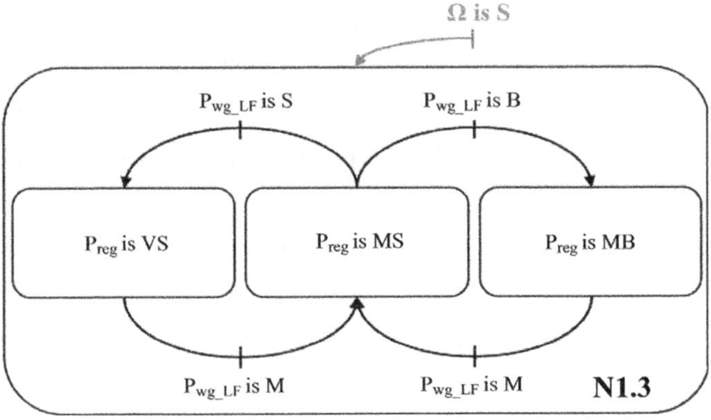

Figure 5.18. *Operational graph of level 1.3*

5.9. Fuzzy rules

The fuzzy rules of the supervisor result from these operational graphs.

Two inputs, Ω and P_{wg_LF}, are subject to fuzzification, each using three linguistic variables; this means that there will be nine fuzzy rules.

<table>
<tr><td rowspan="9">N1</td><td rowspan="3">N1.1</td><td>IF</td><td>Ω is M</td><td>AND IF</td><td>P_{wg_LF} is M</td><td>THEN</td><td>P_{reg} is M</td></tr>
<tr><td>IF</td><td>Ω is M</td><td>AND IF</td><td>P_{wg_LF} is B</td><td>THEN</td><td>P_{reg} is B</td></tr>
<tr><td>IF</td><td>Ω is M</td><td>AND IF</td><td>P_{wg_LF} is S</td><td>THEN</td><td>P_{reg} is S</td></tr>
<tr><td rowspan="3">N1.2</td><td>IF</td><td>Ω is B</td><td>AND IF</td><td>P_{wg_LF} is M</td><td>THEN</td><td>P_{reg} is MB</td></tr>
<tr><td>IF</td><td>Ω is B</td><td>AND IF</td><td>P_{wg_LF} is B</td><td>THEN</td><td>P_{reg} is VB</td></tr>
<tr><td>IF</td><td>Ω is B</td><td>AND IF</td><td>P_{wg_LF} is S</td><td>THEN</td><td>P_{reg} is MS</td></tr>
<tr><td rowspan="3">N1.3</td><td>IF</td><td>Ω is S</td><td>AND IF</td><td>P_{wg_LF} is M</td><td>THEN</td><td>P_{reg} is MS</td></tr>
<tr><td>IF</td><td>Ω is S</td><td>AND IF</td><td>P_{wg_LF} is B</td><td>THEN</td><td>P_{reg} is MB</td></tr>
<tr><td>IF</td><td>Ω is S</td><td>AND IF</td><td>P_{wg_LF} is S</td><td>THEN</td><td>P_{reg} is VS</td></tr>
</table>

Table 5.4. *The nine fuzzy rules of the supervisor*

Table 5.4 presents these rules, grouped according to the sub-levels introduced above.

Note that there can be no reduction of the base number of rules. This is due to the relative simplicity of the case being studied since the fuzzy supervisor has only two inputs. However, this apparent simplicity has to do with the simplification carried out upstream at the time of the supervisor's development. Fuzzy logic has deliberately not been systematically used, and tools such as low-pass filters and PI correctors have been used, thus limiting the number of the fuzzy supervisor's inputs.

5.10. Experimental validation

5.10.1. *Implantation of supervisor*

In order to validate experimentally the concepts proposed, the supervisor developed above has been implanted on "real-time" control cards in a testing platform [CIM 06].

Figure 5.19 represents the nonlinear surface yielded by the fuzzy supervisor. It shows the evolution of the power P_{reg} that must be injected into the electric grid according to the rotation speed of the flywheel, Ω, and of the filtered power produced by the wind generator, P_{wg_LF}. In it, we can see an evolution of P_{reg} in accordance with the rules shown in Table 5.4.

The real-time implementation of this solution is not always directly possible, since the calculation time added to the time needed to take measurements and the time required for the restoration of orders can be incompatible with correct control of the system.

The first solution proposed is a simplification by linearization of the initial fuzzy surface (Figure 5.19) using a mid plan. The equation (in *per units*) of the plan chosen is given by [5.9] and the corresponding surface is shown in Figure 5.20.

$$P_{reg} = 0.63 \cdot P_{wg\_LF} + 0.52 \cdot \Omega - 0.17 \qquad [5.9]$$

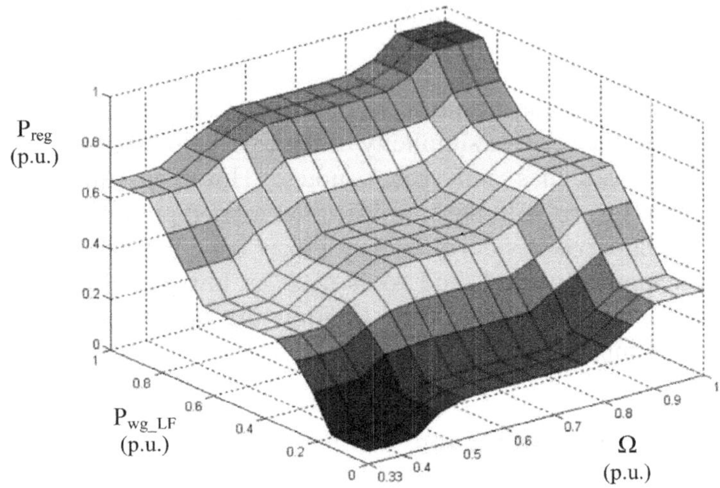

Figure 5.19. *Evolution of the power that must be injected into the grid (P_{reg}) depending on the normalized speed of the flywheel (Ω) and the filtered windpower (P_{wg_LF}). For a color version of the figure, see www.iste.co.uk/robyns/powergrids.zip*

The practical implementation of the simplified supervisor makes it possible to divide the sampling periods of the supervision algorithm implemented in the microprocessor by a factor of 4, in the case of the vectorial control of the asynchronous generator [ROB 07].

Another simplification of the fuzzy supervisor (Figure 5.19) can be done by reproducing extended areas of functioning at constant power of the production system (Figure 5.21). The surface obtained is implemented using simple equations requiring little calculation time.

5.10.2. *Experimental configuration*

The platform used to create experimental tests (Figure 5.22) is structured according to the schema shown in Figure 5.23 [CIM 06, SAU 05].

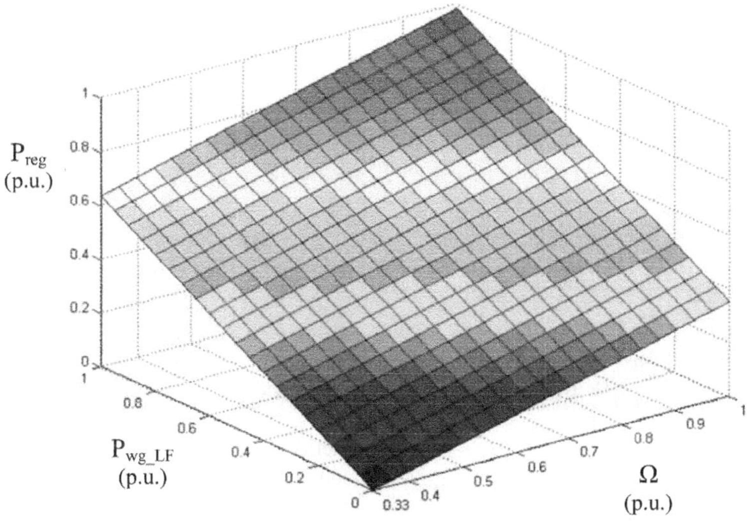

Figure 5.20. *Simplification of supervisor for real-time implementation. For a color version of the figure, see www.iste.co.uk/robyns/powergrids.zip*

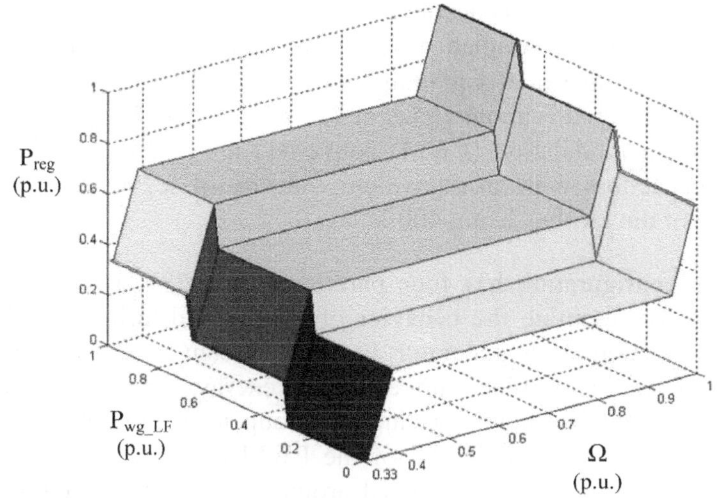

Figure 5.21. *Fuzzy supervisor at constant power. For a color version of the figure, see www.iste.co.uk/robyns/powergrids.zip*

Figure 5.22. *Photo of an experimental platform*

This is a platform with an overall power of around 3 kW. In this configuration, storage is accomplished with the aid of a squirrel-cage induction machine, coupled with a steel flywheel with an inertia of $J = 0.2$ kg.m^2, driven at a maximal speed of 3,000 rpm. The maximal energy thus stored is about 10 kJ, or 2.7 Wh. The DC bus to which the pulse width modulation (PWM) converters are connected is created using capacitors with an equivalent capacity of 2,200 μF, with the voltage of the DC bus being 400 V.

This configuration has four parts (Figure 5.23): a wind turbine emulator representing the behavior of a real wind turbine; a PMSG, whose function as a generator is used to convert the mechanical energy from the turbine into electrical energy; an FESS; and the connection of this energy production group to a three-phase 230 V grid by an L or LC filter. All of the PWM converters used on this platform are identical, constructed around a 1,200 V/50 A insulated gate bipolar transistor (IGBT). The control and measurement interfaces connected to each converter are built, respectively, on a unique model, thus providing a modular quality to this experimental platform.

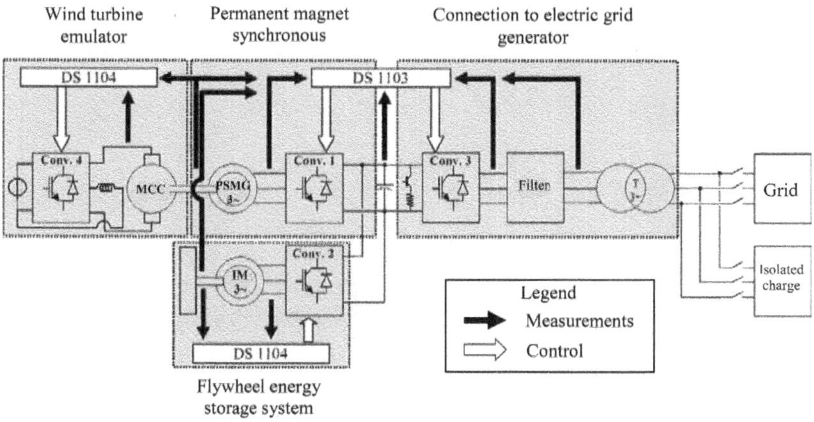

Figure 5.23. *Synoptic schema of an experimental platform*

5.10.3. *Results and analyses*

5.10.3.1. *Smoothed power supervisor*

For these tests, wind measurements recorded on a real site on the northern coast of France are used to create a file yielding the curve shown in Figure 5.24.

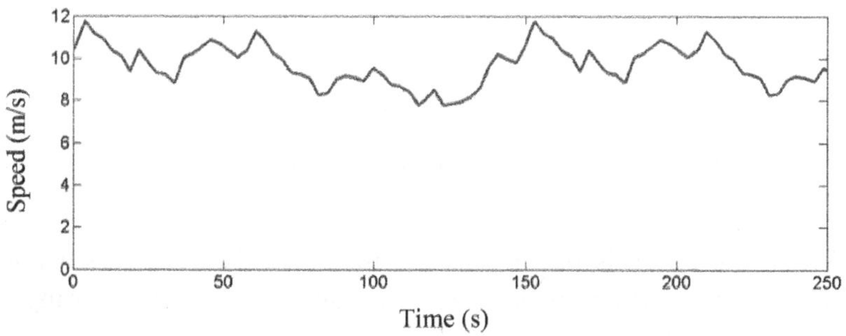

Figure 5.24. *Wind speed*

The drive speed of the wind generator is shown in Figure 5.25.

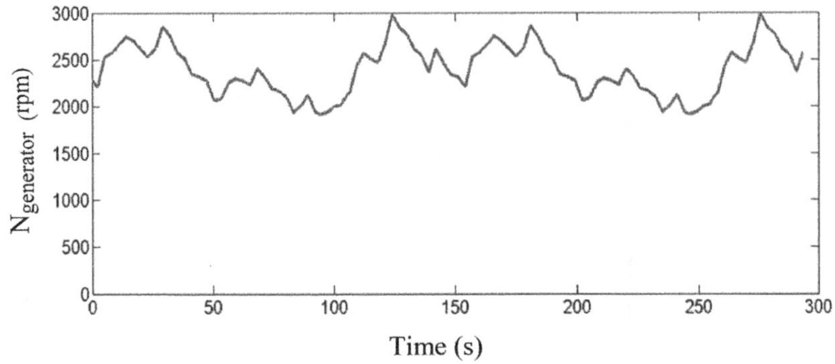

Figure 5.25. *Wind generator speed*

In the absence of a storage system, the wind generator alone injects the highly variable power into the power grid that can be seen in Figure 5.26.

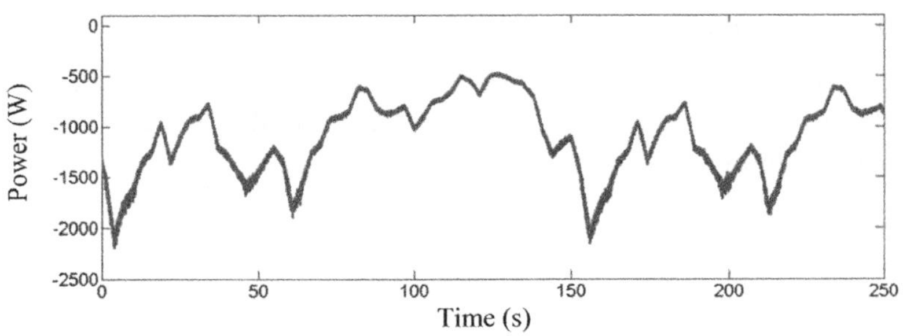

Figure 5.26. *Active power supplied to the grid in the absence of storage*

When the FESS is put in service, the supervisor delivers a power reference, P_{reg}, which is sent to the static converter ensuring the connection with the grid, enabling it to inject the corresponding power into the electric grid (Figure 5.27). As the figure shows, this power is smoothed, which constitutes the fulfillment of the first objective. The development of an indicator could be based here on a frequential analysis of the signal composed of the power injected into the grid.

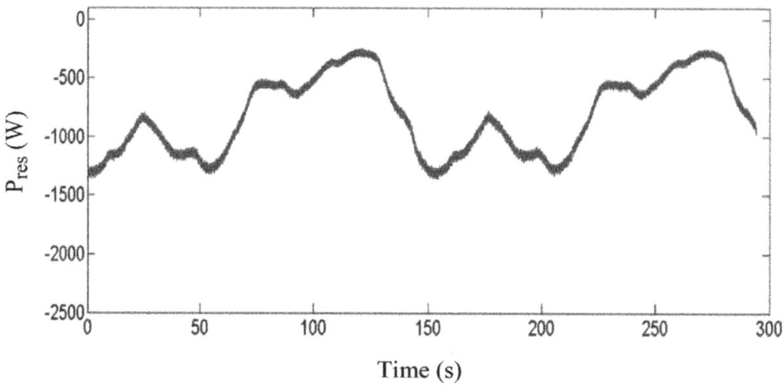

Figure 5.27. *Smoothed active power supplied to electric grid*

During this time, the FESS is requested to absorb or compensate for the power gaps between the wind generator and the power injected into the grid, while maintaining the voltage of the DC bus at its setpoint. The active power absorbed by the induction machine and the rotation speed of the flywheel are shown in Figures 5.28 and 5.29, respectively.

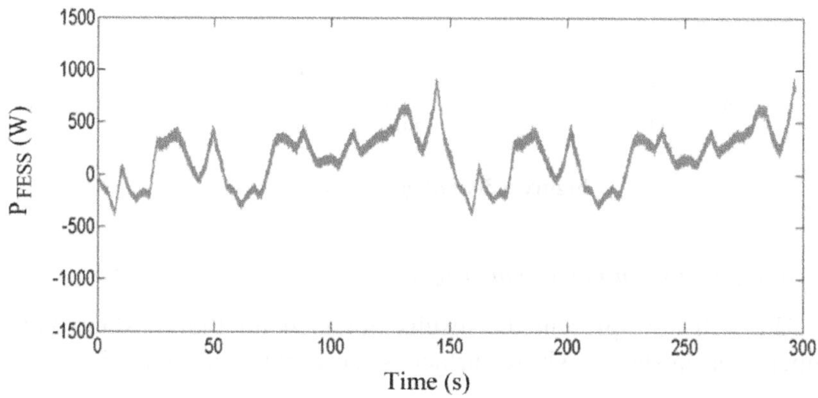

Figure 5.28. *Active power absorbed by an IM*

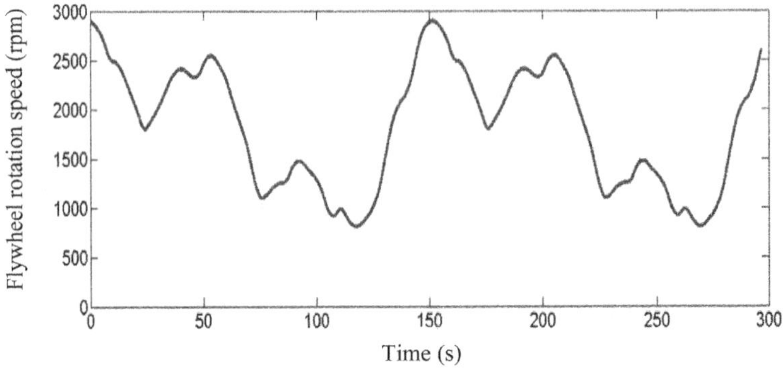

Figure 5.29. *Flywheel speed*

The flywheel speed is maintained between its upper (3,000 rpm) and lower (800 rpm) limits, which fulfills a second objective. The conditions permit then to correctly fulfill the third objective, which is control of the voltage V_{dc} (Figure 5.30).

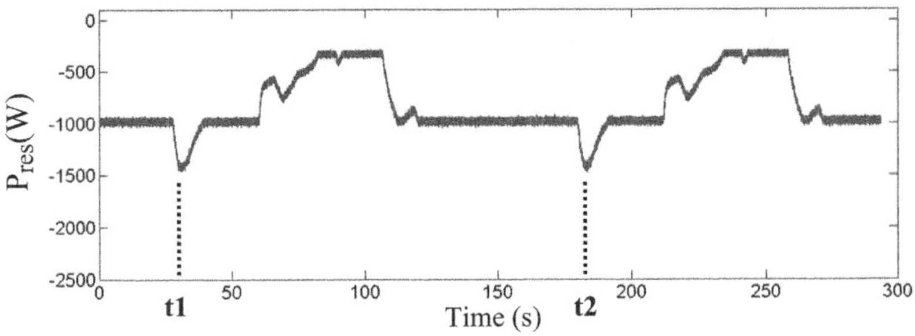

Figure 5.30. *Voltage across DC bus*

5.10.3.2. *Constant power supervisor*

We will now present the results obtained for the constant power supervisor, in stages, whose surface is represented in Figure 5.21.

For a wind speed identical to the one in the previous case (Figure 5.24), the power sent to the grid has the appearance shown in Figure 5.31.

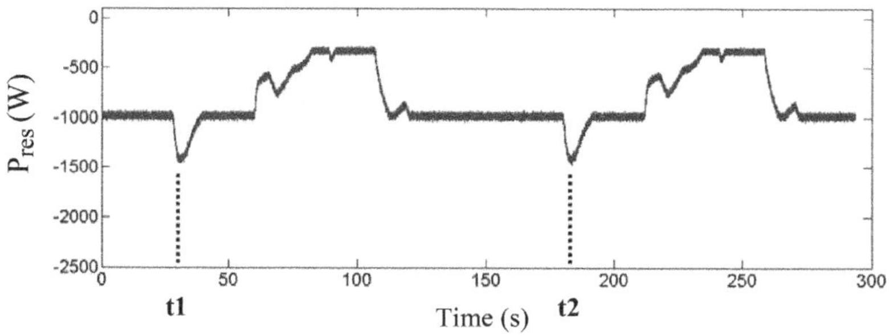

Figure 5.31. *Power sent to grid in the case of a constant power supervisor*

The flywheel speed is represented in Figure 5.32. We can clearly see that the upper and lower limits of the energy storage device's operating phases are reached quite frequently in this case. However, the supervisor, whose objective is to ensure availability of storage, will keep the rotation speed Ω from just reaching the upper and lower saturation zones by releasing more power to the grid when the amount of stored energy is too large (flywheel rotation speed Ω too high – see Figure 5.31, times t_1 and t_2).

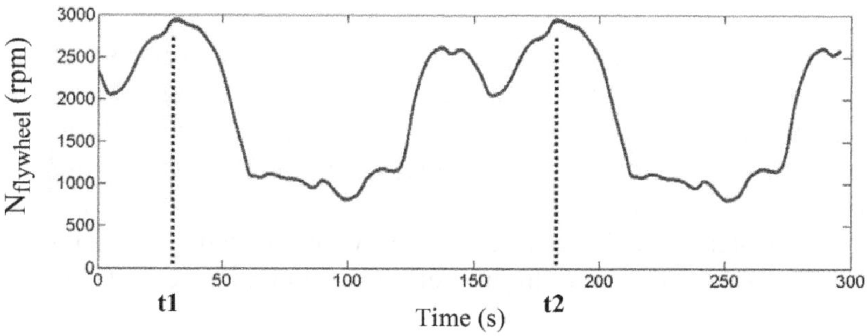

Figure 5.32. *Flywheel rotation speed in the case of a constant power supervisor*

This example illustrates the link that exists between the objectives set for the supervisor and the design of storage capacity. A much higher storage capacity is required for the constant power supervisor since the objective set in terms of the power injected into the grid (in this case, constant in stages) is very binding.

5.11. Conclusion

In this chapter, we have explained the development of an energy supervision strategy within a hybrid power system, using seven precisely identified stages: determination of system specifications, supervisor structure, determination of functional graphs, determination of membership functions of fuzzy supervisor, determination of operational graphs, development of fuzzy rules and definition of indicators.

The application of this methodology to the case of a variable-speed wind turbine combined with flywheel energy storage and connected to the electric grid has been proposed.

While the use of classic tools, such as PI correctors or low-pass filters, enables the production system to satisfactorily fulfill an objective (maintaining DC bus voltage or smoothing the power sent to the grid, respectively), a fuzzy supervisor makes it possible to manage multiple objectives simultaneously (two in this case), and enables storage management via convergence toward a state of charge, thus significantly limiting the risk of storage saturation.

Finally, the implantation of the fuzzy supervisor on an experimental platform has been used to validate its correct functioning. It has been shown that it is possible to inject smoothed active power into the network, whereas the initial power coming from the wind generator is highly variable. This is due to the flywheel storage system, which, when combined with adapted supervision, can also be used to ensure highly effective maintenance of DC bus voltage.

5.12. Bibliography

[BOR 98] BORNE P., RORINOER J., DIEULOT J-Y. *et al.*, *Introduction à la commande floue*, Ed. Technip, Collection: Sciences et Technologies 1998.

[BUH 94] BUHLER H., *Réglage par logique floue,* Presses Polytechniques et Universitaires Romandes, 1994.

[CAR 01] CARDENAS R., PENA R., ASHER G. *et al.*, "Control strategies for enhanced power smoothing in wind energy systems using a flywheel driven by a vector-controlled induction machine", *IEEE Transactions on Industrial Electronics*, vol. 48, pp. 625–635, 2001.

[CIM 05] CIMUCA G., Système inertiel de stockage d'énergie associé à des générateurs éoliens, PhD Thesis, no. 2005-27, Ecole Nationale Supérieure des Arts et Métiers - Centre de Lille, December 2005.

[CIM 06] CIMUCA G., SAUDEMONT C., ROBYNS B. *et al.*, "Control and performance evaluation of a flywheel energy storage system associated to a variable speed wind generator", *IEEE Transactions on Industrial Electronics*, vol. 53, no. 4, pp. 1074–1085, August 2006.

[CIM 10] CIMUCA G., BREBAN S., RADULESCU M. *et al.*, "Design and control strategies of an induction machine-based flywheel energy-storage system associated to a variable-speed wind generator", *IEEE Transactions on Energy Conversion*, vol. 25, no. 2, pp. 526–534, June 2010.

[DAV 06] DAVIGNY A., ROBYNS B., "Fuzzy logic based supervisor of a wind farm including storage system and able to work in islanding mode", *IEEE 32nd Annual Conference on Industrial Electronics*, IECON '06, pp. 4231–4236, 2006.

[EL 09a] EL MOKADEM M., COURTECUISSE V., SAUDEMONT C. *et al.*, "Fuzzy logic supervisor-based primary frequency control experiments of a variable-speed wind generator", *IEEE Transactions on Power Systems*, vol. 24, pp. 407–417, 2009.

[EL 09b] EL MOKADEM M., COURTECUISSE V., SAUDEMONT C. *et al.*, "Experimental study of variable speed wind generator contribution to primary frequency control", *Renewable Energy*, Elsevier, vol. 34, pp. 833–844, 2009.

[JAA 09] JAAFAR A., AKLI C.R., SARENI B. *et al.*, "Sizing and energy management of a hybrid locomotive based on flywheel and accumulators", *IEEE Transactions on Vehicular Technology*, vol. 58, pp. 3947–3958, 2009.

[LEC 03] LECLERCQ L., SAUDEMONT C., ROBYNS B. *et al.*, "Flywheel energy storage system to improve the integration of wind generators into a network", *Electromotion*, vol. 10, pp. 101–106, 2003.

[ROB 07] ROBYNS B., FRANCOIS B., DEGOBERT P. *et al.*, *Vector Control of Induction Machines: Desensitisation and Optimisation Through Fuzzy Logic*, Springer London Ltd, Collection: Power Systems, 2007

[ROB 12] ROBYNS B., DAVIGNY A., BRUNO F. *et al.*, *Production d'énergie électrique à partir des sources renouvelables*, Hermès-Lavoisier, 2012.

[SAU 05] SAUDEMONT C., CIMUCA G., ROBYNS B. *et al.*, "Grid connected or stand-alone real-time variable speed wind generator emulator associated to a flywheel energy storage system", *Power Electronics and Applications European Conference*, p. 10, 2005.

6

Design of a Hybrid Multisource/Multistorage Supervisor

6.1. Introduction

Renewable energy sources such as wind, photovoltaics, small run-of-the-river hydropower and marine sources, such as tides, participate either weakly or not at all in the management of the electrical system to which they are connected (see Chapters 1 and 3). A large quantity of these sources in the energetic mix of the system may cause stability problems [ACK 05, ROB 12b].

These problems are particularly present in the case of low-power island electrical systems or those created following the loss of means of interconnection in the case of continental grids. In these situations, a high penetration of variable renewable energy sources in electricity production systems will be possible only if:

– renewable energy sources (wind, photovoltaic, etc.) are associated with sources whose production is predictable and can therefore be planned and controlled, and/or with storage devices in an integrated management system;

– renewable energy sources participate in the provision of ancillary services such as voltage and frequency control, blackstart, etc.

The improvement of weather forecasting will also contribute to improve the integration of these services into electric grids. The combined management of charges with this type of production will also contribute to a more harmonious integration of renewable sources. This aspect will not be considered in this chapter; an example of this dual management in the case of electric vehicles is discussed in [BOU 13, BOU 14] and [ROB 15].

The association of multiple different energy sources by means of an integrated management system constitutes a virtual multisource power plant. The objective of this chapter is to propose a methodology for the design of supervision of this new kind of production unit [SPR 09, COU 10].

The multisource power plant studied in this chapter is composed of a wind turbine coupled with a predictable and controllable source and with two storage systems with different characteristics. The objectives of supervision will be to follow a reference power while maximizing the use of the renewable resource. Moreover, in the case of frequency variation in the electric grid, the multisource plant must participate in the primary control of this frequency (the principle of which is explained in Chapter 3).

Fuzzy logic-based supervision seems well adapted for the resolution of this type of problem due to:

– the complexity of the system being controlled and the difficulty of obtaining a precise model of it;

– the uncertainty of production of the renewable source;

– the difficulty of quantifying the grid's response to variations in the productions and charges connected to it.

This chapter will show that the method presented in Chapters 1 and 5 makes it possible to:

– avoid the necessity of using precise and complicated models of different sources and storage systems;

– determine the supervisor in a systematic and modular manner;

– ensure progressive transitions between the different operating modes of the hybrid system;

– minimize the number of fuzzy laws (from 540 possible laws to 52 relevant laws in the example considered) and simplify real-time implantation.

In section 6.2, the design methodology for the supervision strategy of the hybrid system will be described. Sections 6.3 and 6.4 use simulations to help illustrate the supervisor performances obtained. In section 6.5, the supervisor design methodology will be tested on various multisource system topologies in order to illustrate its systematic and modular character. The performances of different topologies will be compared using quantitative indicators.

6.2. Methodology for the construction of a supervisor for a hybrid source incorporating windpower

The design methodology of the supervision system applied in this chapter is based on seven stages:

1) determination of system specifications: system characteristics and objectives must be clearly laid out;

2) structure of supervisor: the required supervisor inputs and outputs are determined;

3) determination of "functional graphs": a graphic representation of the operating modes is proposed based on the knowledge of the system;

4) determination of the membership functions of the fuzzy supervisor;

5) determination of "operational graphs": a graphic representation of the fuzzy operating modes is proposed;

6) fuzzy rules: the characteristics of the fuzzy supervisor are extracted from the operational graphs;

7) indicators are used to evaluate the achievement of objectives and to compare different cases (different topologies, variants of supervisor, etc.).

This methodology is developed by applying it to the supervision of a virtual power plant based on renewable energy and including a predictable energy source and two storage systems with different characteristics.

In this study, the sizing of storage systems is not addressed; it is assumed to have been made *a priori* [ROB 12a]. The determination of a real-time supervisor such as the one developed in this chapter may lead to the fine-tuning of the sizing of storage systems with the principal objective of reducing their size.

6.2.1. *Determination of system specifications*

In order to be able to adhere to a reference power and guarantee a reserve stock of power, a windpower plant is connected to a decentralized generator whose production is predictable and can be planned and controlled (e.g. a gas turbine or a diesel generator), and to long-term and short-term storage systems. The whole system is a virtual multisource power plant which will be connected at a point of a power system (Figure 6.1) and will behave like a classic source with regard to the manager of this grid. The principal objectives of a multisource power plant are to:

– supply the reference power set by the grid manager, with the obligations of maximizing the use of renewable resources and minimizing fossil fuel consumption;

– participate in primary frequency control. This objective imposes the obligation of maintaining a power reserve.

In a grid, a gap between the power produced and consumed will be compensated for, by a contribution or a kinetic energy storage in rotating masses of machines. This results in a variation in grid power that will be observed by all of the grid's production units. The lower the total inertia of the rotating masses (and therefore the network is small), the higher the frequency variation created by an imbalance of power will be. The participation of a power generation assembly in frequency control consists of the augmentation of its production in the event of a decrease in frequency, and the reduction of its production in the event of an increase in frequency. This contribution is classically represented by the linear relationship shown in Figure 6.2. The characteristic values of this relationship are its slope and the maximum differential $P_{\max} - P_{ref}$, called the primary reserve ($P_{reserve}$). The size of this stock is defined by the grid manager as a function of economic and technical criteria.

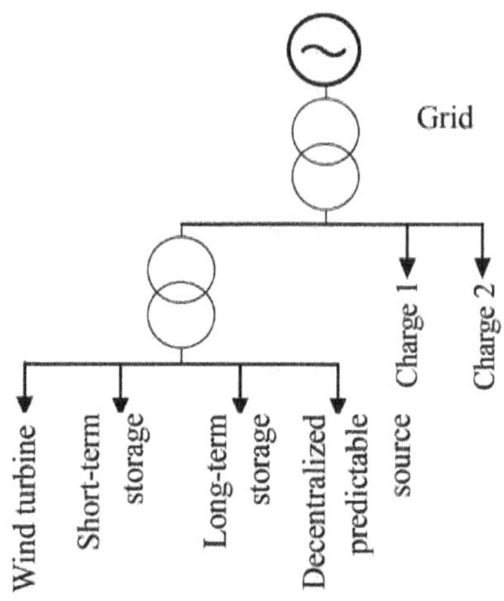

Figure 6.1. *Electrical system studied*

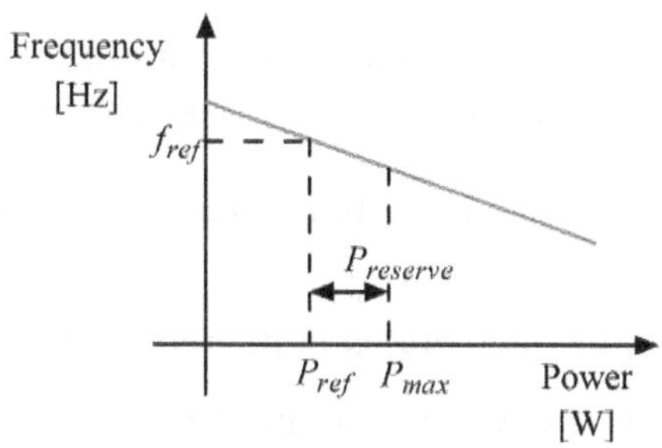

Figure 6.2. *Frequency–power characteristic of a turbo-alternator*

In order to ensure that the wind turbines do not participate in this adjustment for small variations in frequency, a deadband is introduced by grid operators for this characteristic.

Table 6.1 summarizes the objectives, obligations and means of action to be taken into account for the construction of a supervision strategy.

Objectives	Constraints	Means of action
– follow a reference power – contribute to primary frequency control – maximize renewable energy production – minimize fossil fuel consumption – ensure availability of storage systems	– maximum capacity of storage systems – fluctuations in windpower production	– short-term storage reference power – long-term storage reference power – reference power from predictable and controllable source – reference angle of orientation of wind turbine blades

Table 6.1. *Objectives, constraints and means of action to be taken into account when designing a supervision strategy*

6.2.2. *Structure of supervisor*

The structure of a supervisor will be organized so as to achieve the two main objectives defined in the previous section.

Storage systems will play a fundamental role in obtaining these objectives by:

– compensating for fluctuations in the renewable source;

– maintaining a reserve of energy for participating in primary frequency control.

The proper management of these storage systems will be essential to best meet the objectives. Since storage system capacity is limited by technological and economic constraints, however, when these systems are almost completely charged or discharged, the objectives will have to be met by means of other sources:

– by degrading windpower when storage systems are saturated;

– using the predictable source when storage systems are discharged.

At this stage, the inputs required by the supervisor to fulfill its objectives can be defined:

– in order to follow the reference power (P_{ref}), one supervisor input is the difference $\Delta P = P_{ref} - P_{ms}$ between the total power generated by the multisource power plant P_{ms} and the reference power;

– frequency control requires the frequency error (Δf) between the normal-operation frequency of the grid (f_{ref}) and the frequency measured (f_{mes});

– the management of energy reserves requires knowledge of long-term (Niv_{stock_lt}) and short-term (Niv_{stock_ct}) storage levels.

The outputs of the fuzzy supervisor are references imposed on each element of the multisource power plant:

– the reference powers of the storage systems are $P_{ref\_stock\_ct}$ for short-term storage and $P_{ref\_stock\_lt}$ for long-term storage;

– the reference power of the predictable source (P_{sp_ref});

– the reference angle for the calibration of the wind turbine blades β_{ref}. This action is used to degrade the output power of the wind turbine, theoretically extracting its maximum wind energy.

A block diagram of the supervisor is shown in Figure 6.3.

In line with the definition of the two principal objectives, the supervisor is divided into two parts:

– the fuzzy logic-based supervisor, which manages the predictable source (P_{SP_ref}), the wind turbine (β_{ref}) and the storage part ($P_{ref\_stor\_ct\_1}$, $P_{ref\_stock\_lt\_1}$) used to compensate for fluctuations in power from the wind turbine. This supervisor is composed of gains used to normalize inputs (G_1, G_2, G_3) and outputs (G_4, G_5, G_6, G_7);

– the frequency control characteristic will be defined separately from the fuzzy supervisor, and this control will be executed as a matter of priority by the storage systems ($P_{ref\_stock\_ct\_2}$, $P_{ref\_stock\_lt\_2}$), as they constitute the system's energy reserve.

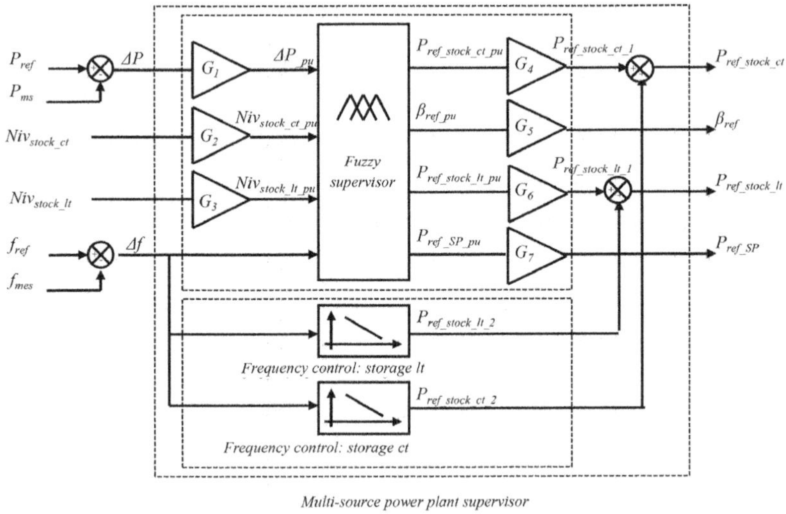

Figure 6.3. *Block diagram of supervisor*

The reference powers of the storage systems are thus the sum of two terms:

$$P_{ref\_stock\_ct} = P_{ref\_stock\_ct\_1} + P_{ref\_stock\_ct\_2} \quad [6.1]$$

$$P_{ref\_stock\_lt} = P_{ref\_stock\_lt\_1} + P_{ref\_stock\_lt\_2} \quad [6.2]$$

It is important to note that in order to guarantee the availability of energy in the storage systems and to ensure that frequency control actions do not compete with the reference control power P_{ref}, storage levels must be managed by the fuzzy supervisor.

6.2.3. *Determination of functional graphs*

The strategy of the fuzzy supervisor of a multisource system can be defined graphically, which has the following advantages:

– a literal expression of the objectives and sub-objectives to be achieved, and the obligations and means of action making it possible to:

- directly establish the fuzzy laws relevant to each operating mode and thus to limit the complexity of the supervisor;

- facilitate exchanges with other disciplinary fields, such as the economy for example, which plays an important role in energy choices;

– a transition between modes determined by the state of certain parts of the system. These states can be described by the fuzzy variables serving as the supervisor inputs, thus enabling smooth transitions between operating modes and the ability of the system to operate in multiple modes simultaneously;

– insofar as fuzzy logic integrates Boolean logic, it can be used to revisit more classic approaches such as Petri and Grafcet diagrams.

The fuzzy logic part of the supervisor shown in Figure 6.3 is represented graphically in Figure 6.4. Operating modes are represented by rectangles with rounded corners and system states by the transitions between these modes.

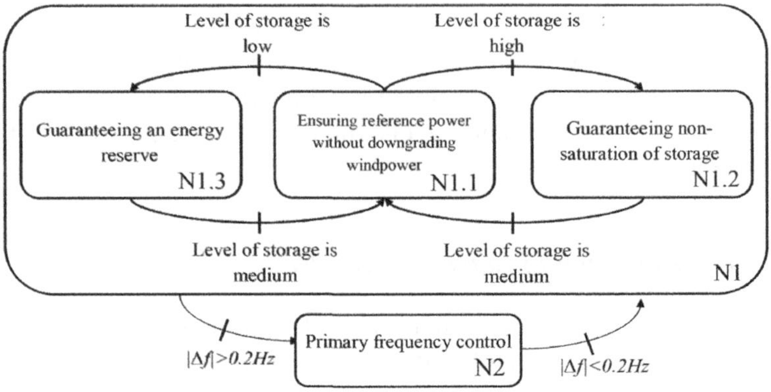

Figure 6.4. *Functional graph of fuzzy logic-based supervisor*

As illustrated in Figure 6.4, the fuzzy logic part of the supervisor is divided into two main operating modes, N1 and N2. The objective of the first operating mode (N1) is to control the output power at the reference power. It is subdivided into three operating submodes. Transitions from one submode to another submode are defined by the state of the storage systems:

– N1.1: *if the storage system level is medium*, the multisource power plant must control the reference power while maximizing the power generated by the wind turbine. The wind turbine therefore functions at its optimal operating point and the storage systems compensate for the difference between the reference power (P_{ref}) and the output power of the multisource power plant (P_{ms}). The functional graph for this operating mode is shown in Figure 6.5;

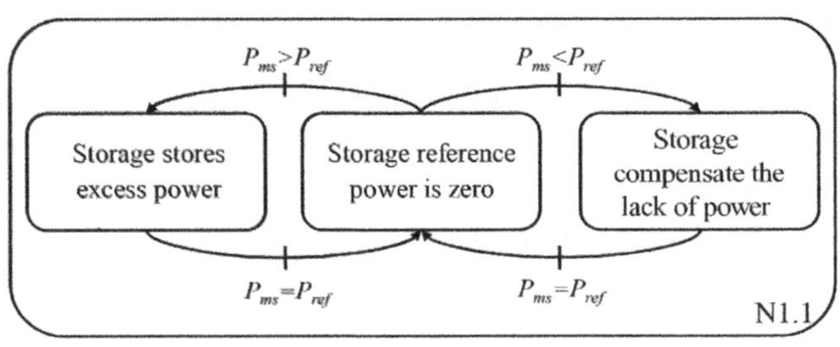

Figure 6.5. *Functional graph of mode N1.1*

– N1.2: *if the storage system level is high*, two actions are executed simultaneously by the supervisor. The first action consists of discharging the storage system in order to participate in primary frequency control by absorbing energy. The second action consists of controlling the power of the multisource power plant at the reference power by modifying the calibration angle of the blades (β). The functional graph for this operating mode is shown in Figure 6.6;

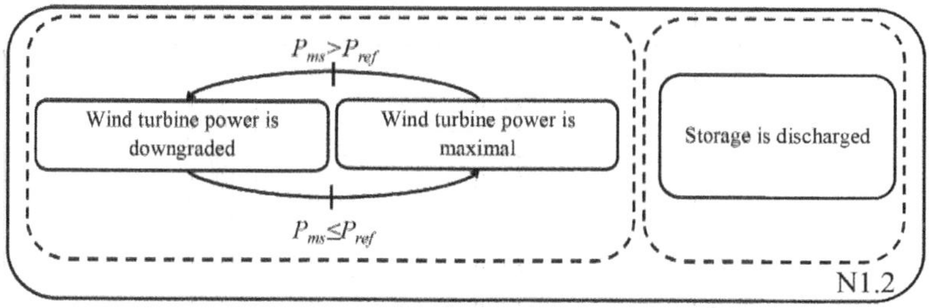

Figure 6.6. *Functional graph of mode N1.2*

– N1.3: *if the storage system level is low*, two actions are executed simultaneously by the supervisor. The first action consists of charging the storage system in order to maintain a reserve stock of energy to participate in primary frequency control. The second action consists of controlling the power of the multisource power plant at the reference power using the predictable source. The functional graph for this operating mode is shown in Figure 6.7.

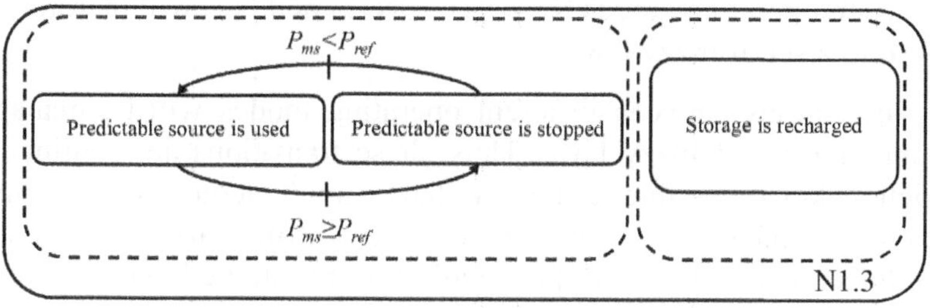

Figure 6.7. *Functional graph of mode N1.3*

The second main operating mode (N2) enables the frequency controller to act via the frequency-power characteristic without entering into conflict with the power controller. In this operating mode, frequency adjustment is prioritized over reference power adjustment. Moreover, in periods of overfrequency and if the storage level is high, the power generated by the wind turbine can be downgraded. Finally, when the predictable source is operational, that is, when the storage level is low, it can also participate in primary frequency regulation. The functional graph for this operating mode is shown in Figure 6.8.

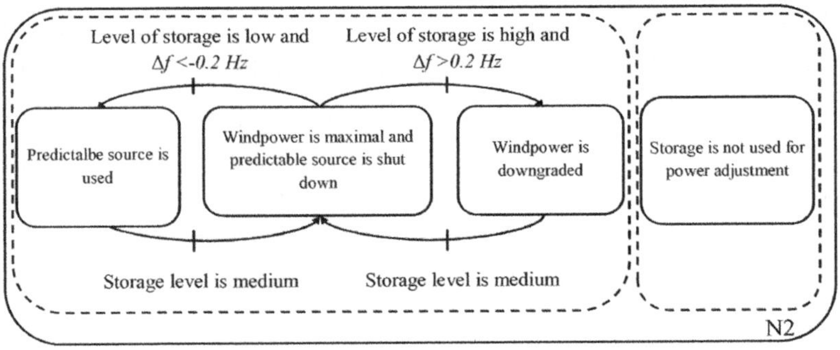

Figure 6.8. *Functional graph of mode N2*

N1.1, N1.2, N1.3 and N2 are the operating modes of the supervisor and are connected to priority objectives, while the transitions (storage level and frequency differential) can be seen as obligations imposed on the system:

– each operating mode (N1.1, N1.2, N1.3 and N2) will therefore be linked to a set of fuzzy laws;

– transitions between different operating modes will be managed by another set of fuzzy laws. Thus, these transitions are continuous, which makes it possible for them to have simultaneous membership in different modes, and thus to tend to fulfill various objectives simultaneously. When multiple conditions are satisfied, multiple fuzzy laws act on the same output. The final value of this output will be the center of gravity of a function determined by fuzzy logic. This approach enables gentle transitions from one operating mode to another.

When two storage systems are considered, the three operating modes N1.1, N1.2 and N1.3 are duplicated as shown in Figure 6.9. N1.1ct, N1.2ct and N1.3ct are linked to the short-term storage system and N1.1lt, N1.2lt and N1.3lt are linked to the long-term storage system. These operating modes are activated simultaneously. Figure 6.9 represents all the operating modes of the fuzzy supervisor as well as the transitions between them.

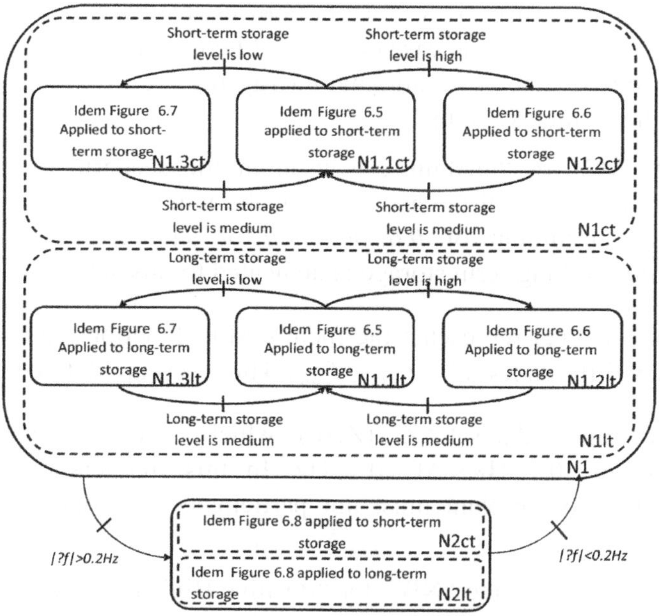

Figure 6.9. *Graphic representation of different operating modes*

6.2.4. *Determination of membership functions*

The next stage in the proposed methodology is the determination of the membership functions of the input and output values of the fuzzy-logic supervisor. The membership functions of the input values will ensure the transitions between different operating modes (Δf, Niv_{Stock_ct} and Niv_{Stock_lt}) or the control of reference values (ΔP and Δf). Since the maximum number of fuzzy laws is a direct function of the number of fuzzy sets considered for each input, we must make sure to minimize the number of these sets. Still for the sake of simplicity, it is advisable to consider symmetric sets without exception. These sets are illustrated in Figure 6.10 for the example being discussed. The membership functions linked to storage levels (Figure 6.10(a) and Figure 6.10(b)) are composed of three levels in line with the three operating submodes (N1.1, N1.2 and N1.3) of the preceding graphic representation:

– sets "S" and "B", for "small" and "big", respectively, ensure the energy reserves necessary for the power plant's contribution to frequency control. In the case considered here, in the event of sub- or super-frequency the same minimal energy reserve of *0.05 per unit* is held in the storage systems;

– set "M", for "medium", is used in the scenario proposed to compensate for the difference between the windpower and reference power. A minimum of *0.6 per unit* of short-term storage and *0.8 per unit* of long-term storage is dedicated to this action.

Figure 6.10(c) represents the membership functions linked to the frequency differences $\Delta f = f_{ref} - f_{mes}$. Three sets are defined:

– one trapezoidal set "Z" (Zero), which is used to introduce a deadband of -0.1 Hz $< \Delta f <$ 0.1 Hz. In this frequency range, the multisource power plant must not participate in primary frequency control;

– two sets "PB" and "NB", for "Positive Big" and "Negative Big", respectively, which are used symmetrically to participate in frequency control.

Figure 6.10(d) represents the membership functions linked to the power differential $\Delta P = P_{ref} - P_{ms}$. Because the minimization of this gap is a major objective, five sets are considered to achieve a compromise between the precision of the power generated and the complexity of the supervisor. The sets are called "NB" (Negative Big), "NM" (Negative Medium), "Z" (Zero), "PM" (Positive Medium) and "PB" (Positive Big).

The membership functions of the output values are illustrated in Figures 6.11(a)–(d), respectively, for the short-term storage reference power, the long-term storage reference power, the reference angle for the orientation of the wind turbine blades and the reference power of the predictable source.

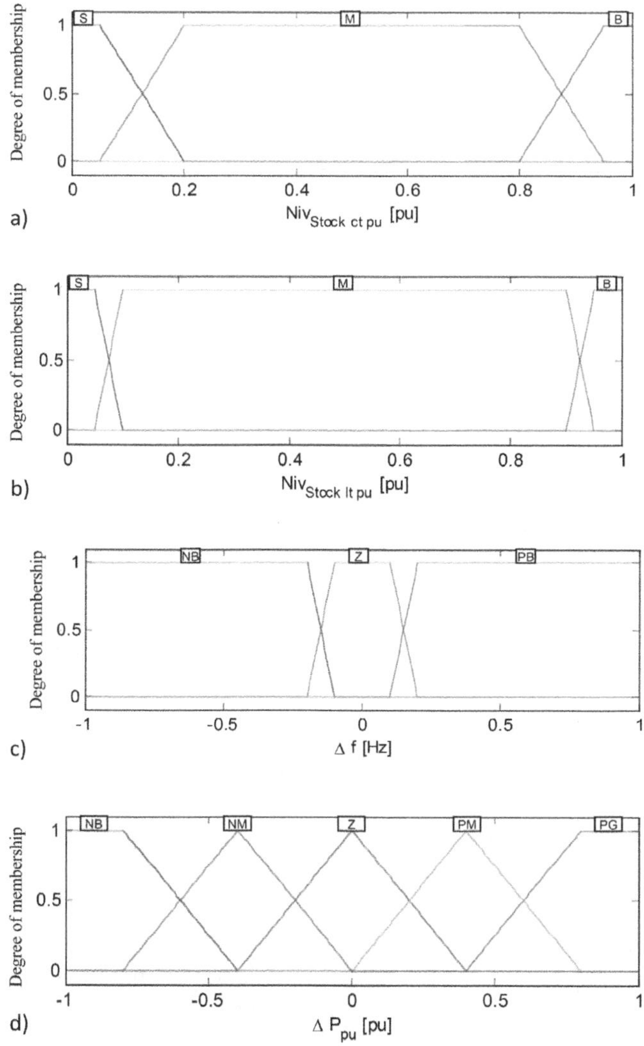

Figure 6.10. *Membership functions of input values: a) short-term storage level, b) long-term storage level, c) frequency differential and d) power differential. For a color version of the figure, see www.iste.co.uk/robyns/powergrids.zip*

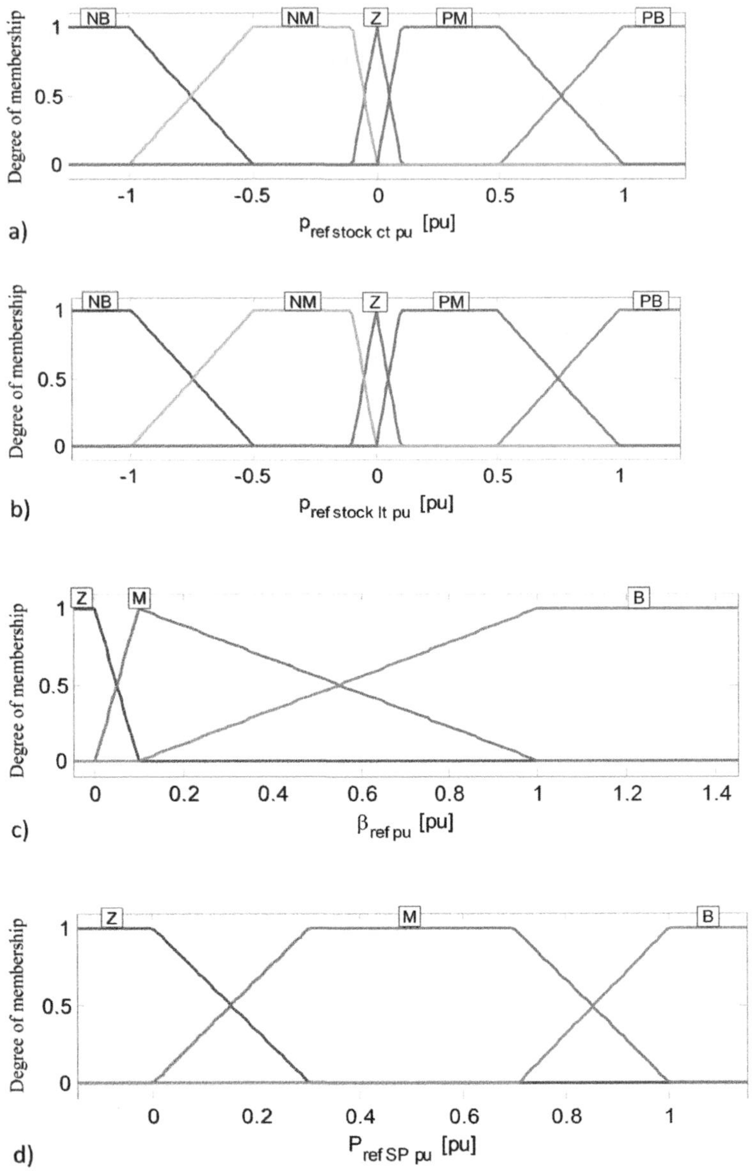

Figure 6.11. *Membership function of output values: a) short-term storage reference power, b) long-term storage reference power, c) reference angle of wind turbine blades and d) reference power of predictable source. For a color version of the figure, see www.iste.co.uk/robyns/powergrids.zip*

Because storage system power levels can be positive or negative, five sets are considered: "NB", "NM", "Z", "PM" and "PB". The choice of these sets is made so that the output value will be included in the interval [–1, 1] (see section 6.5.1). The link between the angle of orientation of the blades and the reduction of windpower is strongly nonlinear; therefore, the membership functions of the calibration angle are chosen to make the relationship between ΔP and the reduction of windpower in operating mode N1.2 more linear. Because the value is always positive [0, 1], three sets ("Z", "M" and "B") are chosen so as to achieve a compromise between precision and complexity. In the example considered, the nature of the predictable source is not taken into account, and any choice of three sets has been made ("Z", "M" and "B") so as to supply a reference power of between [0, 1].

The number of fuzzy laws associated with each output variable is determined by multiplying the numbers of fuzzy sets for each input variable by one another, or 3 × 3 × 3 × 5 = 135. In the case being considered, which includes four output variables, the total number of fuzzy laws possible will be 4 × 135 = 540. Traditionally, these laws are determined using tables. The table associated with each output variable will thus have five dimensions. The methodology proposed enables the determination of the relevant laws. The associated graphic representation has a double advantage; it facilitates the writing of laws without the use of tables and extracts only the laws that are most relevant to the overall functioning of the system.

6.2.5. *Determination of operational graphs*

The schema in Figure 6.9 showed that it is possible to break down the system into a group of subsystems. This breakdown is also used for the determination of fuzzy laws. For this, it is necessary to translate the functional graphs into operational graphs that include the membership functions previously defined. Transitions between operational modes will be described by the membership functions of the input values and the actions of the operational modes will be described by the membership functions of the output values. This approach leads us to the operational graph shown in Figure 6.12. The fuzzy sets of the input variables linked to storage and frequency determine the operational modes. A detail of this figure in the case of submode N1.1ct is presented in Figure 6.13. In this mode, the multisource power plant maximizes windpower (β_{ref_pu} is "Z"), does not use the predictable source ($P_{ref\_SP\_pu}$ is "Z") and controls output power using the short-term storage system; the larger the output power deficit (ΔP_{pu}), the higher the power generated by the storage system; at the same time, the higher the surplus power, the more power is absorbed by the storage system. The fuzzy sets linked to the power differential can thus be used to define the reference power values of the short-term storage system. A similar approach is used for the other operational submodes.

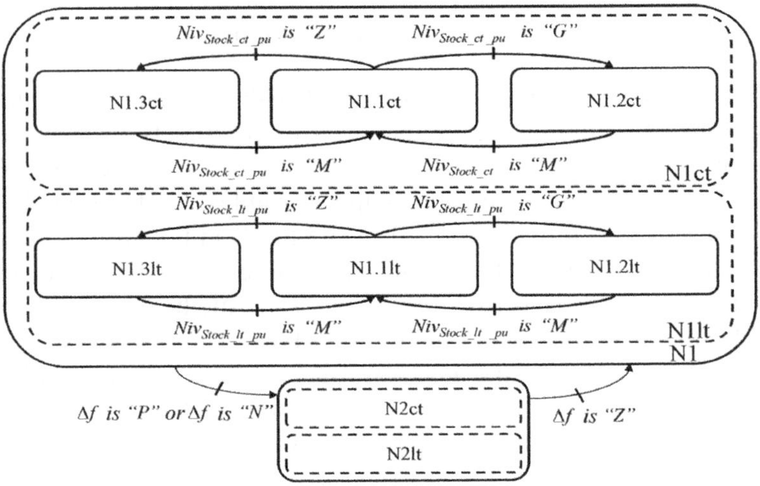

Figure 6.12. *Operational graph of supervisor*

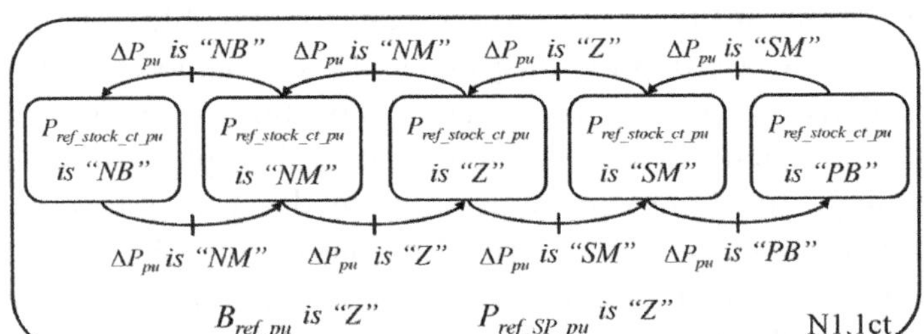

Figure 6.13. *Operational graph of mode N1.1ct*

6.2.6. *Extraction of fuzzy laws*

Based on the diagram shown in Figure 6.13, it is quite simple to extract the fuzzy laws for operational mode N1.1ct. Certain conditions are inherited from the input conditions of modes N1 (**IF** *Δf is "Z"*) and N1.1ct (**IF** *$Niv_{stock\_ct\_pu}$ is "M"*) and a third condition is linked to the fuzzy reasoning operating in this mode:

– **IF** Δf IS Z **AND** $Niv_{stock\_ct\_pu}$ IS M **AND** ΔP_{pu} IS NB **THEN** $P_{ref\_stock\_ct\_pu}$ IS NB;

– **IF** Δf IS Z **AND** $Niv_{stock\_ct\_pu}$ IS M **AND** ΔP_{pu} IS NM **THEN** $P_{ref\_stock\_ct\_pu}$ IS NM;

– **IF** Δf IS Z **AND** $Niv_{stock\_ct\_pu}$ IS M **AND** ΔP_{pu} IS Z **THEN** $P_{ref\_stock\_ct\_pu}$ IS Z;

– **IF** Δf IS Z **AND** $Niv_{stock\_ct\_pu}$ IS M **AND** ΔP_{pu} IS PM **THEN** $P_{ref\_stock\_ct\_pu}$ IS PM;

– **IF** Δf IS Z **AND** $Niv_{stock\_ct\_pu}$ IS M **AND** ΔP_{pu} IS PB **THEN** $P_{ref\_stock\_ct\_pu}$ IS PB.

Finally, two laws determining the output variables linked to the wind turbine and the predictable source are entirely defined by membership in operational mode N1.1ct. These laws are written as:

– **IF** Δf is Z **AND** $Niv_{stock\_ct\_pu}$ is M **THEN** β_{ref_pu} is Z;

– **IF** Δf is Z **AND** $Niv_{stock\_ct\_pu}$ is M **THEN** $P_{ref\_SP\_pu}$ is Z.

Carrying out this operation for all of the modes identified in Figure 6.13, only 52 fuzzy laws are considered instead of the 540 possible laws. A complete list of these laws is given in section 6.5.2. In addition to proposing a systematic construction of the fuzzy supervisor, this methodology enables a significant reduction in the number of laws to be processed, simplifying the real-time implantation of this supervisor.

6.3. Compared performance of different variants of hybrid source

To illustrate the efficiency of the supervisor constructed using the proposed methodology, several simulations of the electromechanical system and its supervisor are presented. A modeling of the various system elements is given in this chapter. However, it is important to note that this modeling effort is not necessary for the development of the supervisor; it is rather used to simulate the system and test the supervisor. The structure of the supervisor is independent of the technologies used to create the virtual power plant, but simulations make it possible to adjust certain supervisor parameters such as the numerical limits of the fuzzy sets of the membership functions.

6.3.1. *Characteristics of simulated system*

6.3.1.1. *Wind*

The variable-speed wind generator simulated in this example is based on a permanent magnet synchronous generator connected to the grid via two back-to-back AC-DC converters.

The turbine is modeled by a relationship linking wind speed to the power that can be extracted from it [ACK 05, ROB 12b]:

$$P_{wind} = \frac{1}{2}\rho A_r C_p(\lambda,\beta) v^3 \qquad [6.3]$$

where ρ is the specific density of the air; A_r is the area swept by the blades; $C_p(\lambda,\beta)$ is the power coefficient; $\lambda = \frac{\Omega_t R_t}{v}$ is the speed ratio, Ω_t is the rotation speed of the turbine, R_t is the radius of the turbine, β is the angle of orientation of the blades; and v is the wind speed. The dependence of the power factor λ and β ($C_p(\lambda,\beta)$) is modeled by an analytical expression proposed in [ACK 05]. The model of the simulated wind turbine is presented in [COU 08a] and [COU 08b].

6.3.1.2. *Predictable source*

A generic predictable source can simply be modeled using a first-order transfer function between the reference power supplied by the supervisor (P_{ref_SP}) and the output power of the source (P_{SP}) [ACK 05]. The equation for this transfer function is given by [6.4], where τ_{SP} is a time constant that depends on the technology chosen:

$$H(s) = \frac{1}{\tau_{SP}\ s + 1} \qquad [6.4]$$

Figure 6.14 represents the model of the predictable source and P_{SP_min} and P_{SP_max} are the minimum and maximum operating powers of the source, respectively.

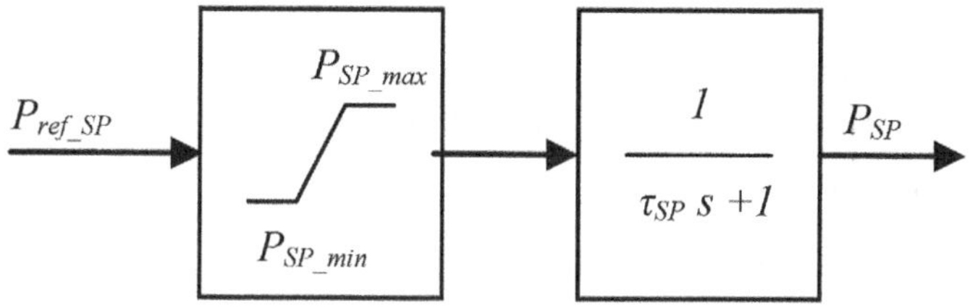

Figure 6.14. *Model of predictable source*

6.3.1.3. *Storage system*

The storage system will be represented by a simplified generic model that is structurally identical for long- and short-term storage systems [ABO 05]. The block schema of this model is shown in Figure 6.15, where P_{ref_stock} is the reference power applied by the supervisor to the storage system, W is the energy stored in the system and P_{stock} is the output power of the system. The characteristic parameters of this model are the maximum charge and discharge powers (P_{chmax} and P_{dchmax}), the charge and discharge time constants (τ_{ch} and τ_{dch}), the efficiency during charge and discharge (η_{ch} and η_{dch}), and the maximum and minimum storage levels (W_{min} and W_{max}). Note that $m_1 = 0$ if $W_{stock} = W_{max}$ where $W_{stock} = 0$; otherwise, $m_1 = 1$.

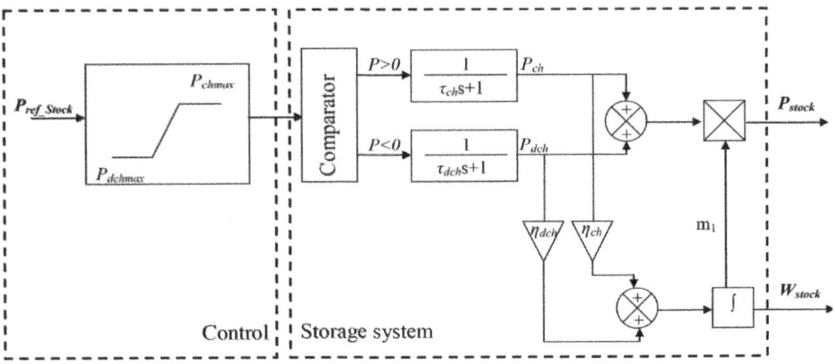

Figure 6.15. *Model of storage system*

6.3.1.4. *Exterior grid*

The exterior grid will be modeled by an equivalent classic production assembly. The elements modeled are:

– the synchronous generator;

– the steam turbine;

– speed control.

Figure 6.16 shows the modeling of the equivalent group when participating in primary and secondary frequency control (a proportional and integral action of speed control) [KUN 97, SAA 99]. T_{CH} and T_{RH} are the time constants of the principal steam input and the heater(s), respectively, and F_{HP} is the fraction of the total power generated by the high-pressure turbine. T_G is the time constant of the speed controller, R is the slope of the droop and K_I is the integral gain of the speed loop. ΔY and ΔP_m are the deviations of the sluice gate and the mechanical power, respectively. ΔP_L is the difference between the planned power from the generator and the power demand of the grid (pu). $H = M/2$ is the kinetic inertia constant (s) and D is the mechanical damping factor.

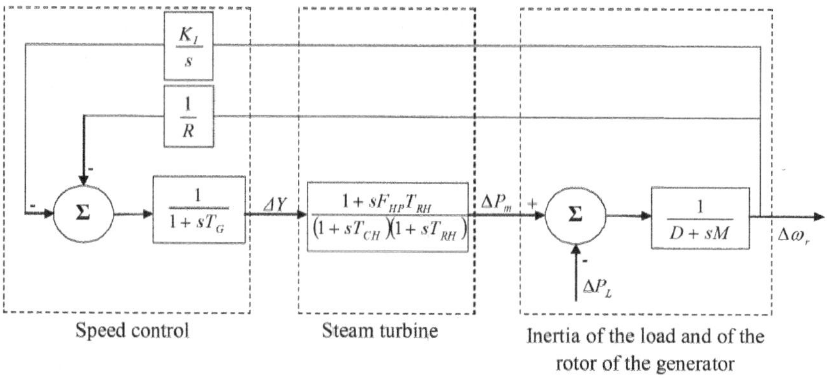

Figure 6.16. *Modeling of equivalent group participating in primary and secondary control*

6.3.2. *Simulations of different hybrid source variants*

To illustrate the advantages of the graphic design method for fuzzy logic-based supervisors, the methodology is applied to different topologies of multisource power plants. For each topology, the graphic representation is derived from Figure 6.9. The supervisor's objectives remain the same as previously, specifically:

– supplying the reference power imposed by the grid manager;

– maximizing the use of renewable resources;

– participating in primary frequency control.

A comparison of the different topologies using quantitative indicators is given at the end of the section.

6.3.2.1. *Simulation of complete multisource power plant (topology A)*

The principal system parameters shown in Figure 6.1 are given in Table 6.2. S_{source} is the apparent power of the grid, P_{load1} is the active power of load 1 and P_{load2} is the active power of load 2. In order to illustrate a high wind penetration rate in an isolated grid (e.g. an island), the power of the grid is considered to be relatively low. R is chosen in accordance with the performance classically demanded of a conventional power group, and K_I is determined so as to obtain a secondary control with a distinctly larger time constant than the primary control. The control of the grid's voltage is assumed to be executed perfectly by the source. In the test scenario proposed, the reference power of the multisource power plant for the period $0h < t < 1h$ is adjusted to *600* kW, which is close to the average windpower. For the period $1h < t < 2h$, the reference power is *400 kW*, that is, lower than the average windpower; and for the final period $2h < t < 3h$, the reference power is higher than the average windpower and is *800* kW. In addition, in order to create frequency variations in the grid, a charge of *800* kW is connected at $t = 0h20$, $t = 1h20$ and $t = 1h40$ and disconnected at $t = 0h40$, $t = 1h40$ and $t = 2h40$.

Predictable source		Grid	
P_{SP}	*750 kW*	S_{source}	*3 MVA*
τ_{SP}	*5 s*	P_{load1}	*800 kW*
		P_{load2}	*800 kW*
		R	*4%*
		K_I	*1*
Short-term storage		**Long-term storage**	
P_{chmax_ct}	*300 kW*	P_{chmax_lt}	*230 kW*
P_{dchmax_ct}	*-300 kW*	P_{chmax_lt}	*-230 kW*
τ_{ch_ct}	*0.5 s*	τ_{ch_lt}	*5 s*
τ_{dch_ct}	*0.5 s*	τ_{dch_lt}	*5 s*
W_{max_ct}	*4.17 kWh*	W_{max_lt}	*417 kWh*
		Windpower	
		P_{wg}	*750 kW*

Table 6.2. *Parameters of simulated grid*

Figure 6.17 shows the results of the simulation; the output power of the wind turbine is represented as a dotted line (Figure 6.17(a)), while the output power of the multisource power plant is shown as a solid line (Figure 6.17(a)). The other subfigures represent, respectively, the frequency of the grid (Figure 6.17(b)), the long-term storage level (Figure 6.17(c)), the short-term storage level (Figure 6.17(d)), the blade calibration angle (Figure 6.17(e)) and the predictable-source power (Figure 6.17(f)).

Figure 6.17(a) shows that the output power of the multisource power plant follows the reference power despite variations in wind and loads. During the connection of loads, the output power is increased in order to reduce the drop in frequency (Figure 6.17(b)), and the opposite phenomenon occurs when a load is disconnected. A closer look at the first frequency drop is shown in Figure 6.17 as a solid line. This figure also shows a comparison of the behavior of the electric system given charge variations with and without the contribution of the multisource power plant. Frequency excursions are reduced by the action of the multisource power plant. In the considered example, the maximum frequency variation is limited to 0.82 Hz, which corresponds to a 40% reduction compared to the situation without a contribution from the multisource power plant.

A simultaneous examination of Figures 6.17(d) and (e) shows that when the level of energy in the short-term storage system is high, blade angle orientation is used to reduce and smoothen the output power from the wind turbine. Similarly, an examination of Figures 6.17(d) and (f) reveals that when the level of energy in the short-term storage system is low, the predictable source is used to compensate for the lack of windpower.

In the simulated case, the predictable source is assumed to be ideal and thus able to generate over the whole range of available power. Depending on the type of source (electric generator, gas micro-turbine, etc.), a minimum power level and obligations related to start-up and shutdown time of this source must be complied with [ALK 09]. Note that this predictable source may also be hydroelectric and may include a water tank [BRE 07, ROB 12b].

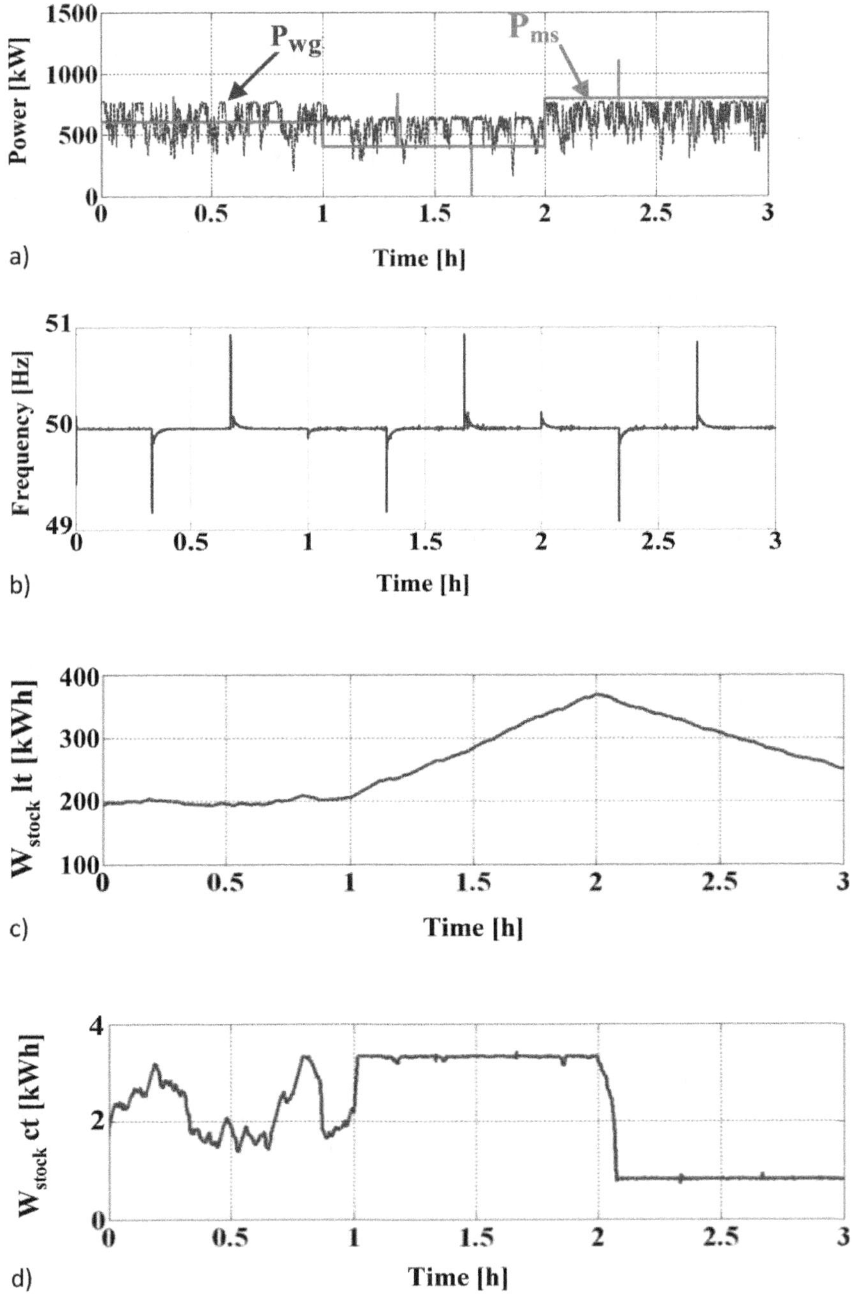
Power [kW]
P_{wg}
P_{ms}
Time [h]
a)
Frequency [Hz]
Time [h]
b)
W_{stock} lt [kWh]
Time [h]
c)
W_{stock} ct [kWh]
Time [h]
d)

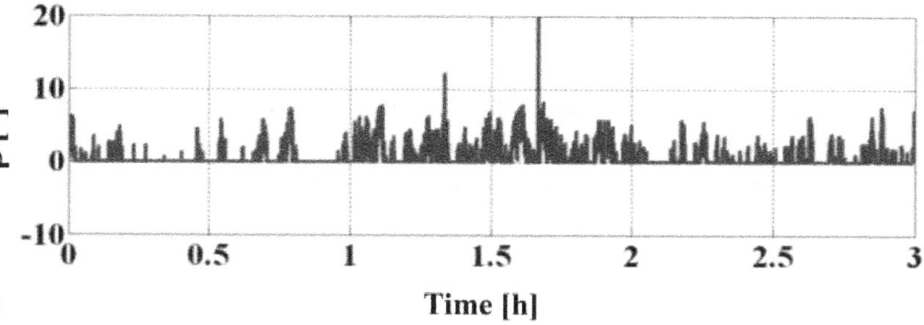

e)

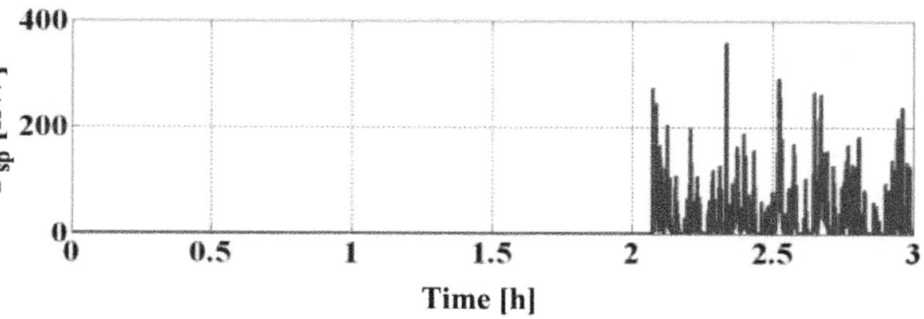

f)

Figure 6.17. *Behavior of multisource power plant for different reference powers: a) windpower and total power of virtual power plant, b) grid frequency, c) long-term storage energy, d) short-term storage energy, e) wind turbine blade orientation angle, f) predictablepower source. For a color version of the figure, see www.iste.co.uk/robyns/powergrids.zip*

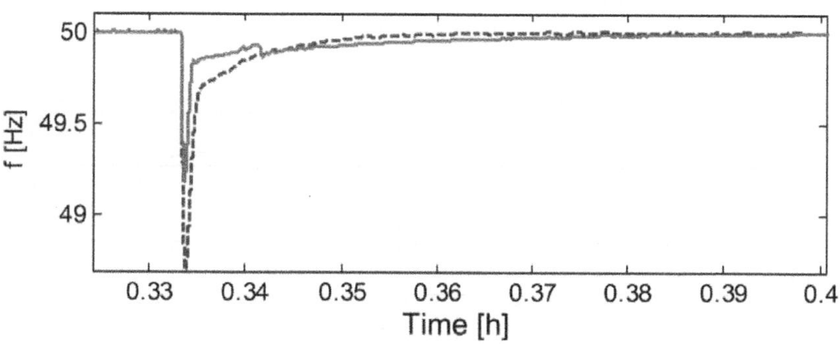

Figure 6.18. *Contribution to frequency control by multisource power plant: frequency variation without contribution from the multisource power plant (dotted line) and with contribution from the multisource power plant (solid line)*

Finally, Figure 6.17(c) shows that the main role of long-term storage is to compensate for low-frequency differences between windpower and the reference power of the multisource power plant. Figure 6.17(d) shows that in normal operation, the storage does not reach its maximum value (4.17 kWh) or its minimum value in order to be able to contribute fully to primary frequency control when necessary, that is, to be able to produce and store energy depending on variations in frequency.

6.3.2.2. *Combination of a wind turbine and a predictable source (topology B)*

The topology considered shown in Figure 6.19 is a multisource power plant composed solely of a wind turbine and a predictable source.

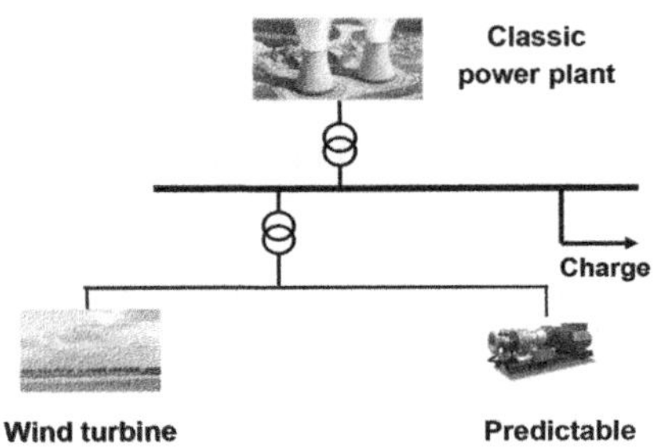

Figure 6.19. *Multisource power plant composed of a wind turbine and a predictable source (topology B)*

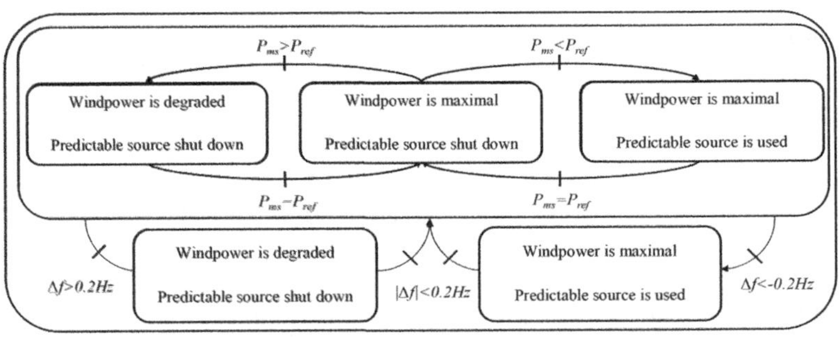

Figure 6.20. *Graphic representation of various operating modes*

Figure 6.20 illustrates the functional graph of the supervisor. This is composed of three operating modes; when the windpower is lower than the reference power, the predictable source supplies the difference. Likewise, when the windpower is higher than the reference power, it must be reduced via the intermediary of the angle of orientation of the wind turbine's blades. Based on this functional graph, we can deduce an operational graph and the relevant fuzzy rules, which number 16. These rules are the ones from sections N1.2ct and N1.3ct of section 6.5.2, which do not draw on the storage reference power and in which the condition having to do with the storage system power level has been removed.

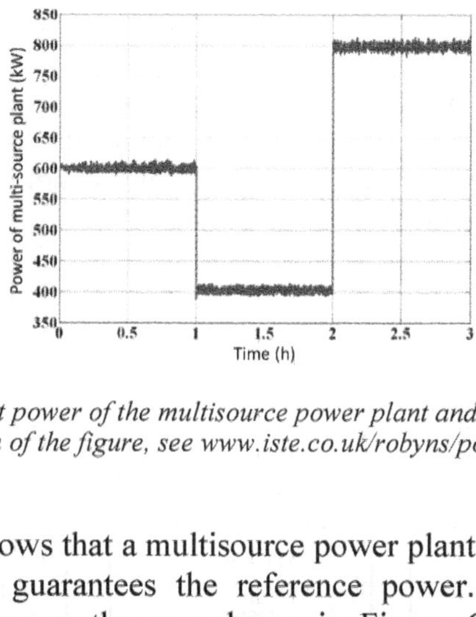

Figure 6.21. *Output power of the multisource power plant and its reference. For a color version of the figure, see www.iste.co.uk/robyns/powergrids.zip*

Figure 6.21 shows that a multisource power plant without a storage system correctly guarantees the reference power. The windpower profile is the same as the one shown in Figure 6.17(a). Note that frequency variations have not been introduced in this scenario. Moreover, in the absence of a storage system in this topology, primary control in the form of a classic droop (Figure 6.2) must be ensured by the predictable source.

6.3.2.3. *Combination of a wind turbine, a predictable source and a short-term storage system (topology C)*

The topology considered is a multisource power plant composed of a wind turbine, a predictable source and a short-term storage system (Figure 6.21).

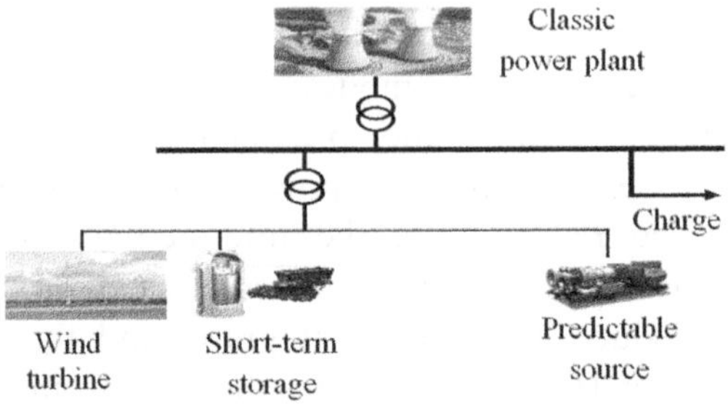

Figure 6.22. *Multisource power plant composed of a wind turbine, a predictable source and a short-term storage system (topology C)*

Figure 6.23 shows the functional graph of the supervisor. It is composed of elements N1.1 and N2 of the supervisor from Figure 6.9. 26 relevant fuzzy rules can be obtained from this graphic representation; these rules are the ones from sections N1ct and N2ct of section 6.5.2, which do not draw on the reference power of the long-term storage system. Figure 6.24 shows that the reference power is properly followed.

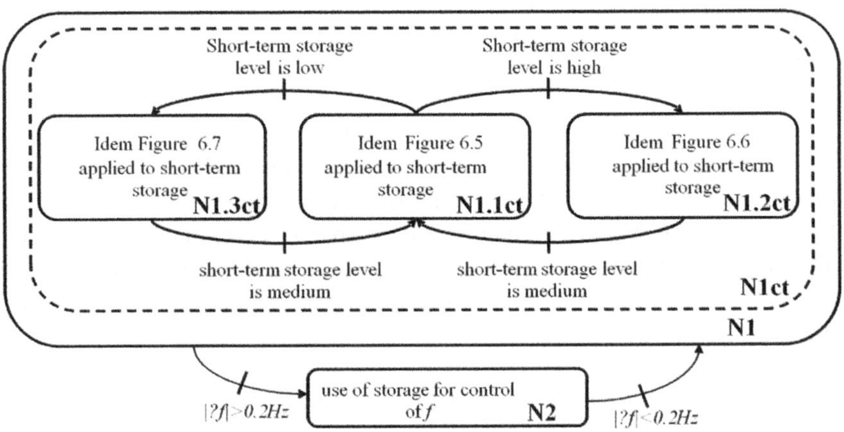

Figure 6.23. *Graphic representation of different operating modes*

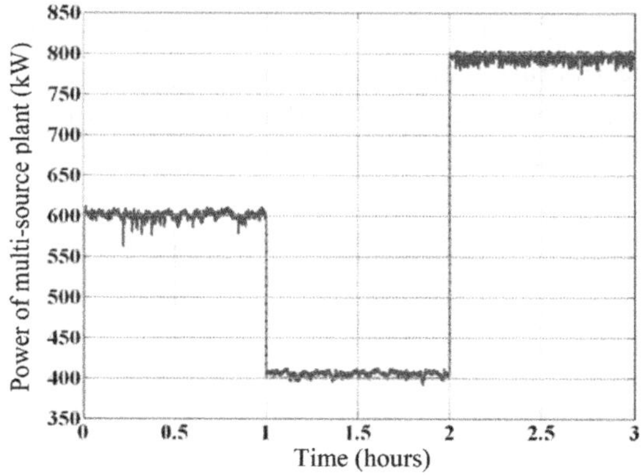

Figure 6.24. *Output power of multisource plant and its reference. For a color version of the figure, see www.iste.co.uk/robyns/powergrids.zip*

6.3.2.4. *Combination of a wind turbine, a short-term storage system and a long-term storage system (topology D)*

The topology considered is a multisource power plant composed of a wind turbine, a short-term storage system and a long-term storage system (Figure 6.25).

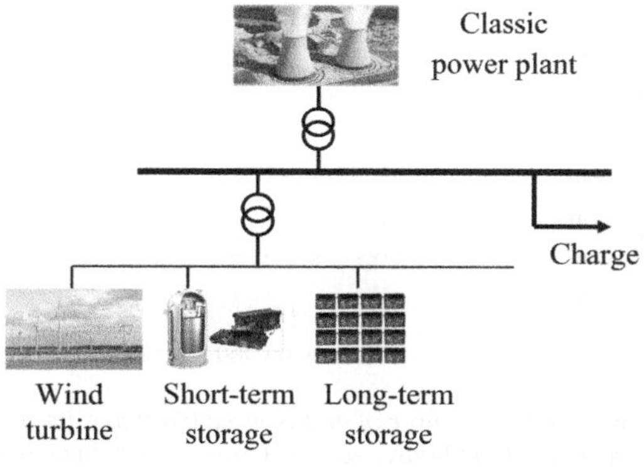

Figure 6.25. *Multisource power plant composed of a wind turbine, a short-term storage system and a long-term storage system (topology D)*

The functional graph of the supervisor is quite similar to the one in Figure 6.9, but the operating modes linked to the predictable source have been removed. Operating mode N1.3, shown in Figure 6.7, is reduced to the operating mode shown in Figure 6.26. The 34 fuzzy rules possible here are the ones from section 6.5.2 that do not draw on the predictable source.

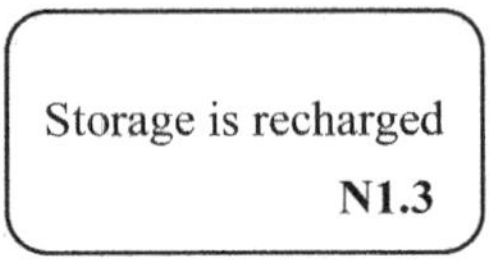

Figure 6.26. *Graphic representation of operating modes N1.3ct and N1.3lt*

Figure 6.27 shows that as long as the reference power is less than or equal to the average windpower, the contribution of the storage systems is enough to control the output power at its reference value. However, this is no longer the case in the third part of the scenario ($2h < t < 3h$), where the reference power is higher than the average windpower (Figure 6.17(a)).

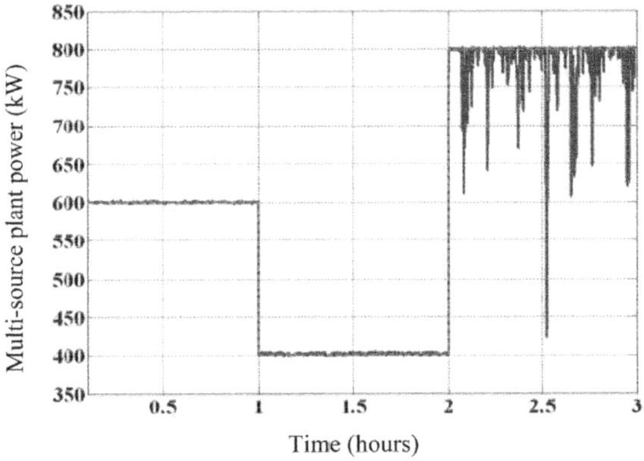

Figure 6.27. *Output power of multisource plant and its reference. For a color version of the figure, see www.iste.co.uk/robyns/powergrids.zip*

6.3.2.5. *Combination of a wind turbine and a short-term storage system (topology E)*

The topology considered is a multisource power plant composed solely of a wind turbine and a short-term storage system (Figure 6.28).

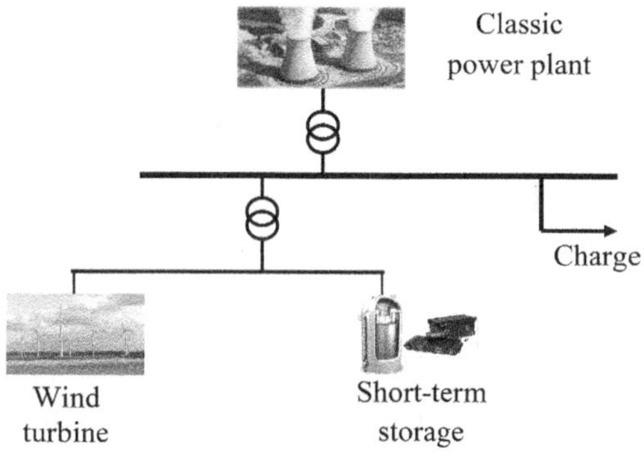

Figure 6.28. *Multisource power plant composed of a wind turbine and a short-term storage system (topology E)*

The functional graph of the supervisor is quite similar to the one shown in Figure 6.23, but the operating modes linked to the predictable source have been removed. Operating mode N1.3, shown in Figure 6.7, is reduced to the operating mode shown in Figure 6.26. The 17 remaining fuzzy rules are the ones from sections N1ct and N2ct of section 6.5.2 that do not draw on either the predictable source or the reference power of the long-term storage system.

Figure 6.29 shows that the reference power is well controlled only when it is lower than the average windpower. When the reference power is higher than the average windpower, regulation can occur only within the limits of the storage system capacity.

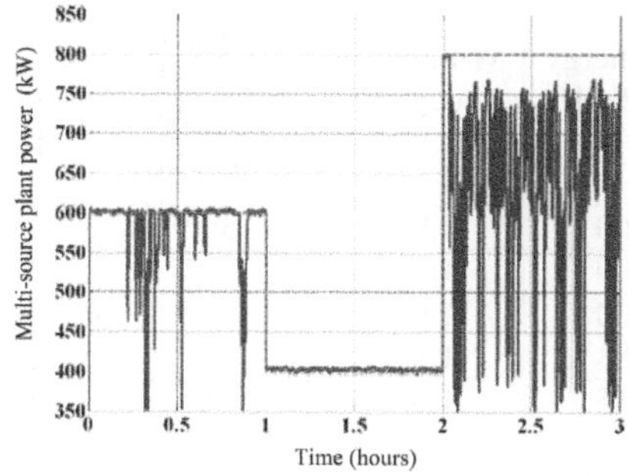

Figure 6.29. *Output power of multisource power plant and its reference. For a color version of the figure, see www.iste.co.uk/robyns/powergrids.zip*

6.3.3. *Comparison of performance of different hybrid sources by means of indicators*

We will now compare five topologies. Topology A corresponds to the multisource power plant composed of all of its elements and simulated in section 6.3.2.1. Table 6.3 offers several comparison criteria in relation to the objectives defined in Table 6.1: total energy supplied by the multisource power plant; energy generated by the wind turbine and by the predictable source as well as the average and maximum error, with error defined as the absolute value of the difference between the reference power and the output power of the multisource power plant. When the predictable source is present (*topologies A, B and C*), the same total energy is sent back to the grid and the average error is low. In all three situations, the presence of storage systems reduces the use of predictable source and increases the use of the renewable source (*topologies A and C*). When the predictable source is not present, the reference power cannot be guaranteed in the presence of a reference power that is higher than the average windpower.

The topology composed of a wind turbine and a short-term storage system is the one that causes the biggest mistake in the monitoring of the reference, with wind energy that is significantly reduced compared to the optimum in terms of the windpower obtained with topologies A and D.

Topology	Total energy supplied [kWh]	Windpower supplied [kWh]	Energy from predictable source [kWh]	Average error [kW]	Maximum error [kW]
A	1,800	1,823	30.9	0.011	18
B	1,800	1,563	237	0.165	13
C	1,800	1,636	163	0.223	38
D	1,783	1,823	0	5.4	376
E	1,635	1,635	0	54.86	511

Table 6.3. *Comparison of different topologies*

6.4. Conclusion

In this chapter, a methodology was presented for the development of a fuzzy logic-based supervisor. This method facilitates the analysis and design of the structure of the supervisor and the fuzzy rules intended to supervise multisource systems. The methodology has been illustrated by the development of the supervision strategy for a multisource power plant composed of a wind turbine, a predictable source and storage systems. This approach enables us to avoid using precise and complex models of the sources in play and makes it possible systematically to determine the supervisor while minimizing the number of rules. The performances of this supervisor have been tested using simulations. Finally, the application of this methodology to different topologies of multisource power plants has been used to illustrate its systematic character and to compare the performance of these power plants. In particular, the positive contribution of a storage system and a predictable source combined with a wind turbine has been illustrated.

6.5. Appendices

6.5.1. *Range of output value variations*

The range of variations of output values is directly linked to the defuzzification of fuzzy output sets. The most widely used method of defuzzification is based on the calculation of the center of gravity of the resulting membership function as follows [6.5]:

$$y^* = \frac{\int_U y.\mu_{res}(y)dy}{\int_U \mu_{res}(y)dy} \qquad [6.5]$$

where U is the universe of discourse of the membership functions of the output values, μ_{res} is the resulting membership function for the output variable y and y^* is the output value. The integral of the denominator gives the surface, while the integral of the numerator corresponds to the moment of this surface.

With this method, the range of variation of y^* will be smaller than its universe of discourse. For this range of variation to be normalized ([–1, 1] or [0, 1]), it is necessary to extend the membership functions outside this range. In the fuzzy supervisor considered, the extreme fuzzy sets (e.g. NB and PB) are, in the range of variation of the output values ([–1, 1] or [0, 1]), triangular in form. It is therefore proposed here to extend these fuzzy sets by a rectangular form outside the normalized range so as to form a trapezoidal fuzzy set as shown in Figure 6.30. Δy_R is the base of the rectangle and Δy_T is the base of the triangle.

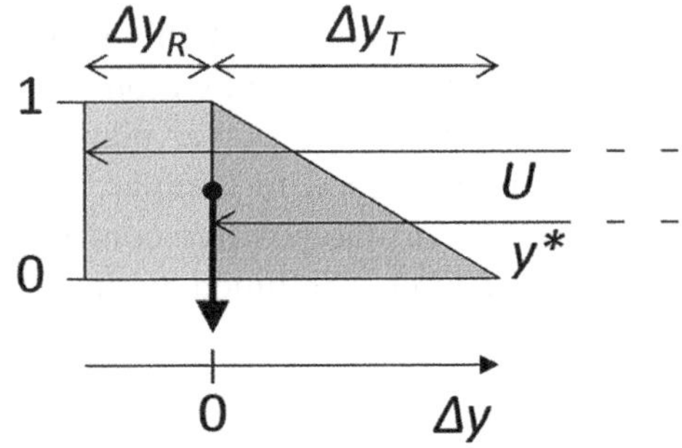

Figure 6.30. *Extreme fuzzy set*

The extreme values of this range of variation will be reached when the resulting membership function is equal to the extreme fuzzy set. For the center of gravity to be located at the interface between the rectangular and triangular part ($\Delta y = 0$), the base of the rectangle Δy_R, and thus the extension of U beyond the normalized range, must be determined by expression [6.6]:

$$\frac{A_R.C_R + A_T.C_T}{A_R + A_T} = \Delta y_R \qquad [6.6]$$

where $A_R = \Delta y_R$ is the area of the rectangle, $C_R = -\Delta y_R/2$ is the abscissa of the center of gravity of the rectangle in the point of reference Δy, $A_T = \Delta y_T/2$ is the area of the triangle and $C_T = \Delta y_T/3$ is the abscissa of the center of gravity of the triangle. By substituting these expressions in equation [6.6], we obtain expression [6.7]:

$$\Delta y_R.\left(-\frac{\Delta y_R}{2}\right)+\frac{\Delta y_T}{3}.\left(\frac{\Delta y_T}{2}\right)=0 \qquad [6.7]$$

After solution, we obtain expression [6.8]:

$$\Delta y_R = \frac{\Delta y_T}{\sqrt{3}} \qquad [6.8]$$

In the example considered of the membership functions of the storage system reference power (Figure 6.11(a)), with the base of the triangle (NB) being 0.5, expression [6.8] shows that the universe of discourse U must be [–1.289, 1.289] for the range of variation of $P_{ref\_stock\_pu}$ to be [–1, 1].

6.5.2. *Fuzzy rules*

			IF Δf is Z **AND** $Niv_{stock\_ct\_pu}$ is M **AND** ΔP_{pu} is NB **THEN** $P_{ref\_Stock\_ct\_pu}$ is NG
			IF Δf is Z **AND** $Niv_{stock\_ct\_pu}$ is M **AND** ΔP_{pu} is PB **THEN** $P_{ref\_Stock\_ct\_pu}$ is PG
			IF Δf is Z **AND** $Niv_{stock\_ct\_pu}$ is M **AND** ΔP_{pu} is NM **THEN** $P_{ref\_Stock\_ct\_pu}$ is NM
		N1.1ct	**IF** Δf is Z **AND** $Niv_{stock\_ct\_pu}$ is M **AND** ΔP_{pu} is PM **THEN** $P_{ref\_Stock\_ct\_pu}$ is PM
			IF Δf is Z **AND** $Niv_{stock\_ct\_pu}$ is M **AND** ΔP_{pu} is Z **THEN** $P_{ref\_Stock\_ct\_pu}$ is Z
N1	**N1ct**		**IF** Δf is Z **AND** $Niv_{stock\_ct\_pu}$ is M **THEN** β_{ref_pu} is Z
			IF Δf is Z **AND** $Niv_{stock\_ct\_pu}$ is M **THEN** $P_{ref\_SP\_pu}$ is Z
			IF Δf is Z **AND** $Niv_{stock\_ct\_pu}$ is B **AND** ΔP_{pu} is Z **THEN** β_{ref_pu} is Z
			IF Δf is Z **AND** $Niv_{stock\_ct\_pu}$ is B **AND** ΔP_{pu} is NM **THEN** β_{ref_pu} is Z
		N1.2ct	**IF** Δf is Z **AND** $Niv_{stock\_ct\_pu}$ is B **AND** ΔP_{pu} is NB **THEN** β_{ref_pu} is Z
			IF Δf is Z **AND** $Niv_{stock\_ct\_pu}$ is B **AND** ΔP_{pu} is PM **THEN** β_{ref_pu} is M
			IF Δf is Z **AND** $Niv_{stock\_ct\_pu}$ is B **AND** ΔP_{pu} is PB

			THEN β_{ref_pu} is G
			IF Δf is Z **AND** $Niv_{stock\_ct\_pu}$ is B **THEN** $P_{ref\_SP\_pu}$ is Z
			IF Δf is Z **AND** $Niv_{stock\_ct\_pu}$ is B **THEN** $P_{ref\_Stock\_ct\_pu}$ is NM
		N1.3ct	**IF** Δf is Z **AND** $Niv_{stock\_ct\_pu}$ is S **AND** ΔP_{pu} is NB **THEN** $P_{ref\_SP\_pu}$ is B
			IF Δf is Z **AND** $Niv_{stock\_ct\_pu}$ is S **AND** ΔP_{pu} is NM **THEN** $P_{ref\_SP\_pu}$ is M
			IF Δf is Z **AND** $Niv_{stock\_ct\_pu}$ is S **AND** ΔP_{pu} is PB **THEN** $P_{ref\_SP\_pu}$ is Z
			IF Δf is Z **AND** $Niv_{stock\_ct\_pu}$ is S **AND** ΔP_{pu} is PM **THEN** $P_{ref\_SP\_pu}$ is Z
			IF Δf is Z **AND** $Niv_{stock\_ct\_pu}$ is S **AND** ΔP_{pu} is Z **THEN** $P_{ref\_SP\_pu}$ is Z
			IF Δf is Z **AND** $Niv_{stock\_ct\_pu}$ is S **THEN** β_{ref_pu} is Z
			IF Δf is Z **AND** $Niv_{stock\_ct\_pu}$ is S **THEN** $P_{ref\_Stock\_ct\_pu}$ is PM
	N1lt	**N1.1lt**	**IF** Δf is Z **AND** $Niv_{stock\_lt\_pu}$ is M **AND** ΔP_{pu} is NB **THEN** $P_{ref\_Stock\_lt\_pu\_pu}$ is NB
			IF Δf is Z **AND** $Niv_{stock\_lt\_pu}$ is M **AND** ΔP_{pu} is PB **THEN** $P_{ref\_Stock\_lt\_pu}$ is PB
			IF Δf is Z **AND** $Niv_{stock\_lt\_pu}$ is M **AND** ΔP_{pu} is NM **THEN** $P_{ref\_Stock\_lt\_pu}$ is NM
			IF Δf is Z **AND** $Niv_{stock\_lt\_pu}$ is M **AND** ΔP_{pu} is PM **THEN** $P_{ref\_Stock\_lt\_pu}$ is PM
			IF Δf is Z **AND** $Niv_{stock\_lt\_pu}$ is M **AND** ΔP_{pu} is Z **THEN** $P_{ref\_Stock\_lt\_pu}$ is Z
			IF Δf is Z **AND** $Niv_{stock\_lt\_pu}$ is M **THEN** β_{ref_pu} is Z
			IF Δf is Z **AND** $Niv_{stock\_lt\_pu}$ is M **THEN** $P_{ref\_SP\_pu}$ is Z
		N1.2lt	**IF** Δf is Z **AND** $Niv_{stock\_lt\_pu}$ is B **AND** ΔP_{pu} is NB **THEN** β_{ref_pu} is Z
			IF Δf is Z **AND** $Niv_{stock\_lt\_pu}$ is B **AND** ΔP_{pu} is Z **THEN** β_{ref_pu} is Z
			IF Δf is Z **AND** $Niv_{stock\_lt\_pu}$ is B **AND** ΔP_{pu} is NZ **THEN** β_{ref_pu} is Z
			IF Δf is Z **AND** $Niv_{stock\_lt\_pu}$ is B **AND** ΔP_{pu} is PM **THEN** β_{ref_pu} is M
			IF Δf is Z **AND** $Niv_{stock\_lt\_pu}$ is B **AND** ΔP_{pu} is PB **THEN** β_{ref} is B
			IF Δf is Z **AND** $Niv_{stock\_lt\_pu}$ is B **THEN** $P_{ref\_SP\_pu}$ is Z
			IF Δf is Z **AND** $Niv_{stock\_lt\_pu}$ is B **THEN** $P_{ref\_Stock\_lt\_pu}$ is NM

		N1.3lt	**IF** Δf is Z **AND** $Niv_{stock\_lt\_pu}$ is S **AND** ΔP_{pu} is NB **THEN** $P_{ref\_SP\_pu}$ is B
			IF Δf is Z **AND** $Niv_{stock\_lt\_pu}$ is S **AND** ΔP_{pu} is NM **THEN** P_{ref_SP} is M
			IF Δf is Z **AND** $Niv_{stock\_lt\_pu}$ is S **AND** ΔP_{pu} is PB **THEN** $P_{ref\_SP\_pu}$ is Z
			IF Δf is Z **AND** $Niv_{stock\_lt\_pu}$ is S **AND** ΔP_{pu} is PM **THEN** $P_{ref\_SP\_pu}$ is Z
			IF Δf is Z **AND** $Niv_{stock\_lt\_pu}$ is S **AND** ΔP_{pu} is Z **THEN** $P_{ref\_SP\_pu}$ is Z
			IF Δf is Z **AND** $Niv_{stock\_lt\_pu}$ is S **THEN** β_{ref_pu} is Z
			IF Δf is Z **AND** $Niv_{stock\_lt\_pu}$ is S **THEN** $P_{ref\_Stock\_lt\_pu}$ is PM
N2	N2ct		**IF** Δf is PB **OU** Δf is N **THEN** $P_{ref\_Stock\_ct\_pu}$ is Z
			IF Δf is NB **AND** $Niv_{stock\_ct\_pu}$ is B **THEN** β_{ref_pu} is Z
			IF Δf is PB **AND** $Niv_{stock\_ct\_pu}$ is B **THEN** β_{ref_pu} is M
			IF Δf is PB **AND** $Niv_{stock\_ct\_pu}$ is S **THEN** $P_{ref\_SP\_pu}$ is Z
			IF Δf is NB **AND** $Niv_{stock\_ct\_pu}$ is S **THEN** $P_{ref\_SP\_pu}$ is M
	N2lt		**IF** Δf is PB **OU** Δf is N **THEN** $P_{ref\_Stock\_lt\_pu}$ is Z
			IF Δf is NB **AND** $Niv_{stock\_lt\_pu}$ is B **THEN** β_{ref_pu} is Z
			IF Δf is PB **AND** $Niv_{stock\_lt\_pu}$ is B **THEN** β_{ref_pu} is M
			IF Δf is PB **AND** $Niv_{stock\_lt\_pu}$ is S **THEN** $P_{ref\_SP\_pu}$ is Z
			IF Δf is NB **AND** $Niv_{stock\_lt\_pu}$ is S **THEN** $P_{ref\_SP\_pu}$ is M

6.6. Bibliography

[ABO 05] ABOU CHACRA F., Valorisation et optimisation du stockage d'énergie dans un réseau d'énergie électrique, PhD Thesis, University of Paris-Sud 11, 2005.

[ACK 05] ACKERMANN T., *Wind Power in Power Systems*, John Wiley & Sons, New York, 2005.

[ALK 09] ALKHALIL F., DEGOBERT P., COLAS F. *et al.*, "Fuel consumption optimization of a multimachines microgrid by secant method combined with IPPD table", *International Conference on Renewable Energy and Power Quality*, ICREPQ '09, Valence, 15–17 April 2009.

[BOU 13] BOUALLAGA A., MERDASSI A., DAVIGNY A. *et al.*, "Minimization of energy transmission cost and CO2 emissions using coordination of electric vehicle and wind power (W2V)", *IEEE PES International Conference POWERTECH 2013*, Grenoble, 16–20 June 2013.

[BOU 14] BOUALLAGA A., DAVIGNY A., MERDASSI A. *et al.*, "Optimization of fuzzy supervisor for electric vehicle load in distribution grid", *11th International Conference on Modeling and Simulation of Electric Machines, Converters and Systems*, ELECTRIMACS '14, Valence, Spain, 20–22 May 2014.

[BRE 07] BREBAN S., NASSER M., ANSEL A. *et al.*, "Variable speed small hydro power plant connected to AC grid or isolated loads", *Journal European Power Electronics*, vol. 17, pp. 29–36, no. 4, 2007.

[COU 08a] COURTECUISSE V., MOKADEM M.E., ROBYNS B. *et al.*, "Supervision par logique floue d'un système éolien à vitesse variable en vue de contribuer au réglage primaire de fréquence", *Revue internationale de génie électrique*, nos. 4–5, pp. 423–453, July–October 2008.

[COU 08b] COURTECUISSE V., SPROOTEN J., ROBYNS B. *et al.*, "Experiment of a wind generator participation to frequency control", *Journal European Power Electronics*, vol. 18, no. 3, pp. 14–24, 2008.

[COU 10] COURTECUISSE V., SPROOTEN J., ROBYNS B. *et al.*, "Methodology to build fuzzy logic based supervision of hybrid renewable energy systems", *Mathematics and Computers in Simulation*, Elsevier, vol. 81, pp. 208–224, October 2010.

[KUN 97] KUNDUR P., *Power System Stability and Control*, WCB/McGraw-Hill, 1997.

[ROB12a] ROBOAM X., *Integrated Design by Optimization of Electrical Energy Systems*, ISTE, London and John & Wiley Sons, New York, 2012.

[ROB 12b] ROBYNS B., DAVIGNY A., BRUNO F. *et al.*, *Electric power generation from renewable sources*, ISTE, London and John & Wiley Sons, New York,, 2012.

[ROB 15] ROBYNS B., SAUDEMONT C., HISSEL D. *et al.*, "Management and valorization of energy storage in transportation and buildings", ISTE-Wiley, forthcoming, 2015.

[SAA 99] SAADAT H., *Power System Analysis*, WCB/McGraw-Hill, 1999.

[SPR 09] SPROOTEN J., COURTECUISSE V., ROBYNS B. *et al.*, "Méthodologie de développement de superviseurs à logique floue de centrales multi sources à base d'énergie renouvelable", *European Journal of Electrical Engineering*, vol. 12, nos. 5–6, pp. 553–583, 2009.

7

Management and Economic Enhancement of Adiabatic Compressed-Air Energy Storage Incorporated into a Power Grid

7.1. Introduction

The technique of storing energy in the form of compressed air in underground caverns hybridized with a gas turbine (or CAES, for compressed air energy storage system) is an alternative, in the range of several hundred MW power generation, to the turbine-pump stations installed in mountainous regions, for which most of the available sites are already in use. However, this technique requires large investments and has the disadvantage of low energy efficiency, less than 50%, which is difficult to define precisely due to the fact that natural gas is burned during the generating phase. However, another CAES technology is in the process of emerging: adiabatic CAES, or A-CAES. Here, a thermal storage stage is added in order to recover the heat energy released when air is compressed; this energy is then reused to heat the air during expansion. The expected efficiency is thus around 65% (see section 2.5.3 in Chapter 2). Figure 7.1 illustrates the principle of this kind of storage.

The objective of this chapter is to analyze the economic value-enhancement and interest, and thus the uses of medium- and high-power (15–30 MW and 100–300 MW) A-CAES storage devices for a power grid.

Existing storage technologies are considered to have reached the limit of their profitability in the current electric system when only traditional economic enhancement related to supply and demand is taken into account. In this context, maximization of the services provided and the profitability of the storage system can be achieved only through the localization and optimal dimensioning of the storage system in the electric grid, and via its optimal temporal management (see Chapter 3). This supervision is available on various time scales: long-term (the day ahead), medium-term (between 1 h and 30 min in advance) and short-term (real time).

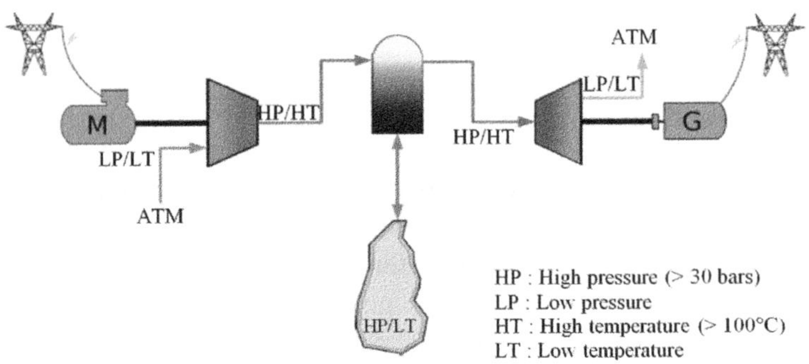

Figure 7.1. *Adiabatic compressed-air storage in a cavern (source: EDF)*

This study involves the development of a real-time storage supervision strategy intended to maximize services provided and profitability following predimensioning and prelocalization [DO 14]. A method for constructing a fuzzy logic-based supervisor is applied according to the methodology described in Chapter 5. The 14-node IEEE test grid described in Figure 7.19 has been chosen as an example to illustrate the application of the method. Three variants of supervisor will be compared: a supervisor limited to traditional economic enhancement based on supply and demand and planned the day ahead, the proposed real-time supervisor based on fuzzy logic and finally a

Boolean variant of this latter supervisor. Simulation results show an economic storage gain that becomes considerably more interesting when it participates in ancillary services requiring real-time management.

7.2. Services provided by storage

7.2.1. *Storage planning*

An initial economic enhancement of a storage system is accomplished through the electricity market, based on the purchase/sale mechanism, in the day ahead.

In order to ensure the economic enhancement of the storage system, several additional services may be offered to stakeholders in the electric system. In the context of this study, only required services and services with a significant economic advantage will be considered: frequency control, congestion management and renewable production support services.

7.2.2. *Frequency control*

This major service is broken down into three services: primary control, secondary control and tertiary adjustment; however, only the first two services are mandatory [ROB 12] (see Chapter 3).

As part of primary frequency control, if the production power is higher than 40 MW, it must be able to maintain a minimal power reserve of 2.5% of the power installed. In addition, service must be maintained at the minimum for 15 min when the frequency variation is between 49.8 and 50.2 Hz, and this reserve must be completely liberated in less than 30 s. This adjustment is carried out by means of a droop between power and frequency (explained in Chapter 3).

Secondary control is mandatory for plants with nominal power higher than 120 MW, and the reserve must be equal to at least 4.5% of the nominal power. Its deployment must begin within 30 s of the incident and end within 7 min, and in the event of a significant frequency variation this service must be maintained for as long as is necessary.

7.2.3. *Congestion management*

The treatment of congestion consists of relieving a transmission line overloaded by the transit of power. Because of their reversibility, storage devices are of interest as means of resolving congestion (see Chapter 3).

The long-term treatment of congestion implies that the transmission system operator (TSO) reinforces the network by building additional power lines. However, construction costs are very high, and the process may take more than 10 years. In view of this, the use of a storage system by the TSO can postpone costly investments and speed up the strengthening of the grid [VER 09, VER 11].

7.2.4. *Guarantee of variable renewable production*

A high renewable production insertion rate can cause rapid reversals of power fluxes due to the random nature of the primary source (wind or sun). This phenomenon has already appeared on the lines interconnecting Germany and the countries bordering it, due to the high concentration of wind energy in northern Germany [ACK 05]. The errors in predicting windpower on a regional scale vary on average between 3% for a 1-h prediction and 7% for a prediction of 72 h in advance, which is quite satisfactory for controlling the supply-demand balance. However, it is around 15% for a wind turbine park, with a significant disparity depending on local topography [ACK 05].

Currently, the additional reserves necessary to compensate for these uncertainties are taken on by classic generators; however, eventually, due to the growing participation of sources that are difficult to predict in the energetic mix, managers of variable renewable power plants will have to comply with a power profile produced the day ahead, as with classic power plants. Storage can then provide these producers with a service to guarantee production, the principle of which is based on the storage of power produced that exceeds forecasts, and on compensation for power shortages when these occur. Renewable producers can thus avoid penalty charges by using storage systems that will cost their owners less than the cost of these charges.

7.3. Supervision strategy

7.3.1. *Methodology*

The integration of decentralized sources and loads in future smart grids requires the development of economic enhanced energy storage via a mutualization of services and multi-objective supervision adapted to the integration of multi-source, multi-charge systems. One challenge for the development of these supervision strategies is the random behavior of the systems concerned, whose temporal horizons can be quite short (dynamic stresses) or very long (the seasonal nature of renewable sources). Classic (or explicit) optimization methods are difficult to apply in real time, and are not easily exploitable when the temporal horizon of a study must cover a whole year, and when systems whose states are dependent on time (such as storage) must be taken into account.

On the other hand, implicit methods using artificial intelligence tools such as fuzzy logic are well suited to the management of "complex" systems dependent on values or states that are difficult to predict and thus are not well known (wind, sun, state of grid, consumption, etc.). A supervisor design methodology based on fuzzy logic for the management of hybrid energy-production systems was introduced in Chapters 5 and 6. This methodology does not require mathematical models as it is based on system expertise represented by fuzzy rules. Inputs may be random, and supervision may target multiple objectives simultaneously. Transitions between operating modes are progressive since they are determined by fuzzy variables. Finally, this methodology enables storage management via convergence toward a charge level (state of charge (SOC)) and control of complexity with a view to real-time implementation.

The construction of a fuzzy supervisor is based on the methodology presented in Chapter 5 and is organized into eight stages:

– determination of system specifications: objectives, constraints and means of action are identified;

– structure of supervisor: required inputs and outputs are determined;

– determination of "functional graphs": a graphic representation of operating modes is proposed, based on system knowledge;

– determination of membership functions of fuzzy variables;

– determination of "operational graphs": a graphic representation of fuzzy operating modes is proposed;

– extraction of fuzzy rules from operational graphs;

– determination of indicators used to evaluate the achievement of objectives;

– optimization of supervisor parameters, for example, by means of experimental and genetic algorithm plans.

7.3.2. *Objectives, constraints and means of actions*

The structure of a supervisor is organized so as to achieve the three main objectives defined in Table 7.1. The constraints of supervision are also presented.

A long-term supervisor is assumed to have defined, in the day ahead, a power reference for storage taking into account the electricity market and the planning of the grid. The objectives of a multi-objective fuzzy supervisor can be divided into three groups: economic objectives, required services and additional services.

7.3.3. *Supervisor structure*

One supervisor input will correspond to each objective. The supervisor structure is shown in Figure 7.2. Four inputs are identified:

– To ensure the availability of storage for frequency control and other services, the state of charge of the storage system must be considered as an input (SoC).

– The second input is the reference for non-mandatory or additional services $P_{service}$, which is the sum of the congestion management service $P_{congestion}$ and the renewable production guarantee service $P_{guarantee}$.

– The third supervisor input will be the storage power plan P_{plan}. Storage planning is done the day ahead based on the price curve and needs of the grid.

– Frequency control requires direct action on the storage system due to the dynamics and mandatory character of this service; this is why it acts directly on the output reference of the fuzzy supervisor to generate the final reference $P_{setpoint}$.

Objectives	Constraints	Means of action
– Maximize economic gain while respecting the planned power – Ensure primary frequency control – Supply additional ancillary services – Ensure availability of storage	– Storage limits – Transit capacity of line – Variations of windpower	– Power storage reference

Table 7.1. *Objectives, constraints and means of action*

In Figure 7.2, K1, K2, K3 and K4 are the per-unit adaptation coefficients of the input and output variables.

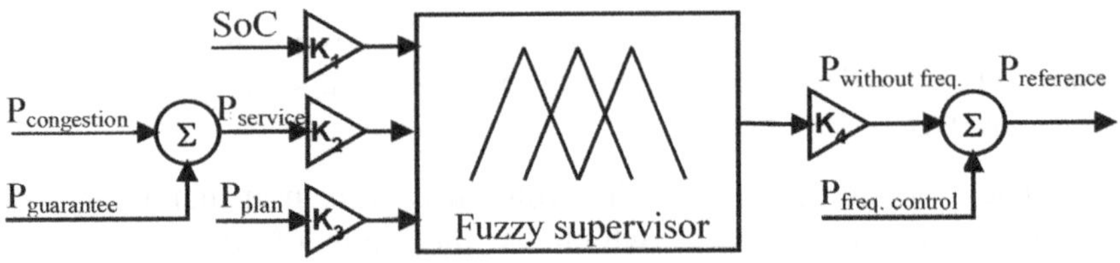

Figure 7.2. *Inputs and outputs of real-time supervisor*

7.3.4. *Determination of functional graphs*

Operating modes are represented graphically in Figure 7.3 by the rectangles with rounded corners, and system states by the transitions between these modes.

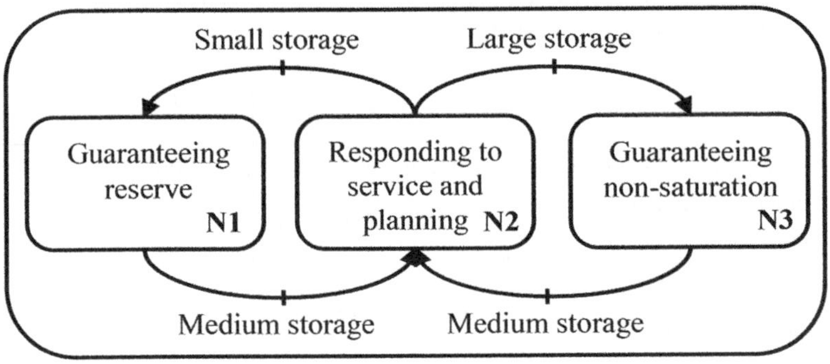

Figure 7.3. *Functional graph of fuzzy logic-based supervisor*

Basing itself on the level of storage (SOC), the functioning of the storage system can be divided into three operating modes. Transitions from one mode to another are defined by the SOC. A negative reference power corresponds to a storage charge, and conversely a positive reference power corresponds to a storage discharge.

N1 (Figure 7.4): if the storage system level is small, the storage system is no longer able to discharge in order to preserve the quantity of energy necessary for frequency control. If the storage system is requested to charge, it will operate in a way favorable to charging.

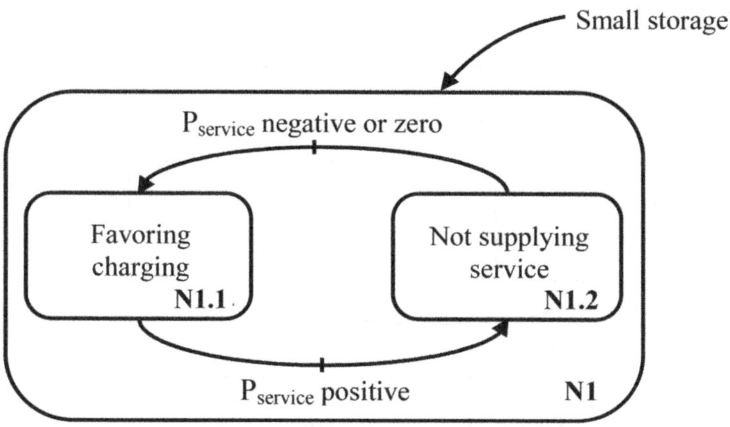

Figure 7.4. *Functional graph of mode N1*

N2 (Figure 7.5): if the storage system level medium-sized, it can follow the additional services reference. If no reference is given, the storage system will respect the planned power in order to maximize economic gain.

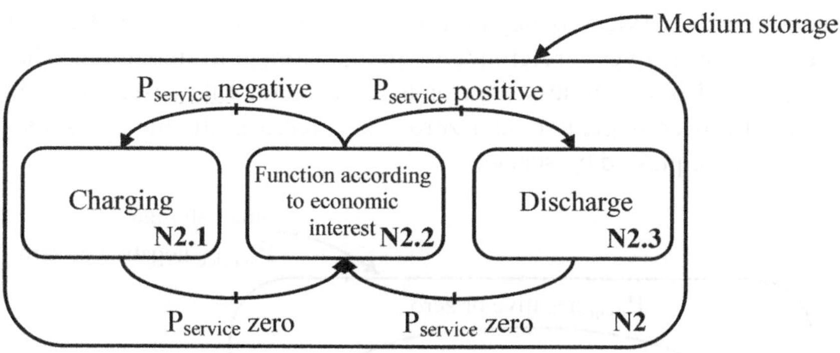

Figure 7.5. *Functional graph of mode N2*

N3 (Figure 7.6): if the storage system level is large, the storage system is unable to charge in order to avoid saturation. If it is requested to discharge, it will operate in a way favorable to discharging.

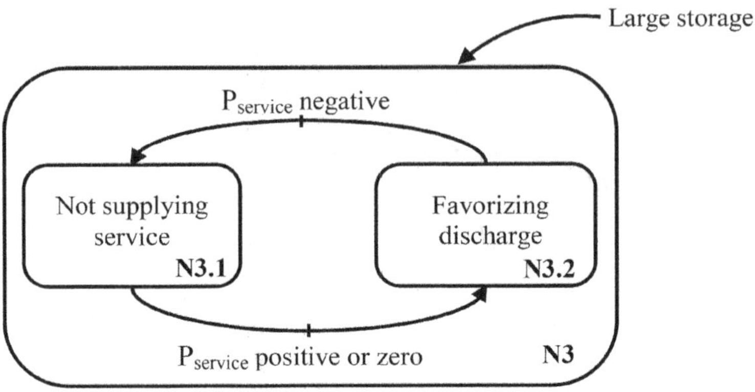

Figure 7.6. *Functional graph of mode N3*

For modes N1.2 and N3.1, one fuzzy rule can be established directly for each block. They are independent of the last input, the planned power. For the rest, more detailed functional graphs must be given.

N1.1 (Figure 7.7): the storage level is small and the power requested by services is negative or zero. To favor storage charging, the supervisor gives a reference to charge storage to maximum when the planned power is less than or equal to zero. Likewise, when the planned power is greater than zero, the reference attempts to satisfy the power requested by services.

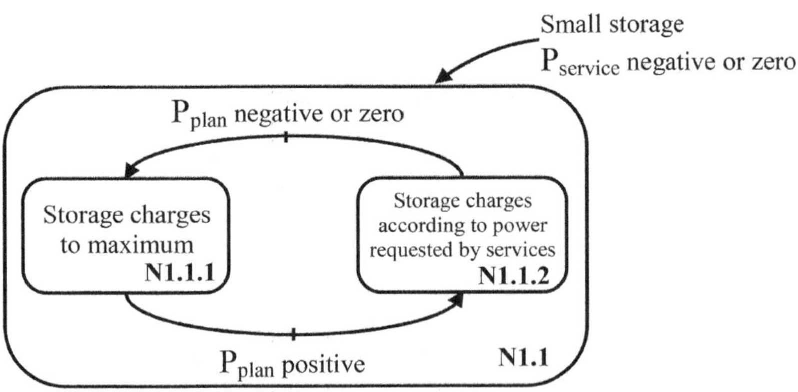

Figure 7.7. *Functional graph of mode N1.1*

N2.1 (Figure 7.8): the storage level is medium and the power requested by services is negative. In this mode, the supervisor gives the reference to charge storage to maximum when the planned power is negative. If necessary, it gives a reference to fulfill the service request.

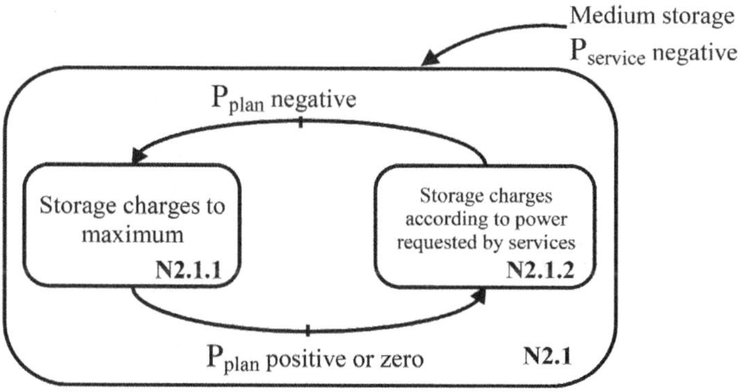

Figure 7.8. *Functional graph of mode N2.1*

N2.2 (Figure 7.9): the storage level is medium and the power requested by services is zero. In this mode, the supervisor gives a reference according to an economic interest, determined by planning: if planned power is negative, then storage charges to the maximum; if planned power is positive, then storage discharges to the maximum; in other words, if planned power is zero, storage will be put into standby mode.

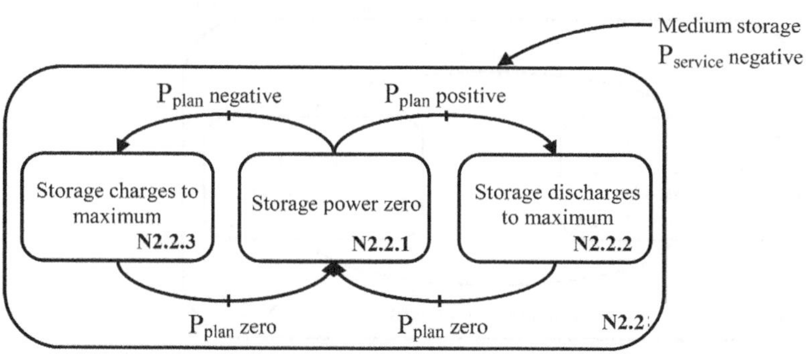

Figure 7.9. *Functional graph of mode N2.2*

N2.3 (Figure 7.10): the storage level is medium and the power requested by services is positive. In this mode, the supervisor gives a reference to charge storage to maximum when the planned power is lower than zero. If necessary, it gives a reference to fulfill service requests.

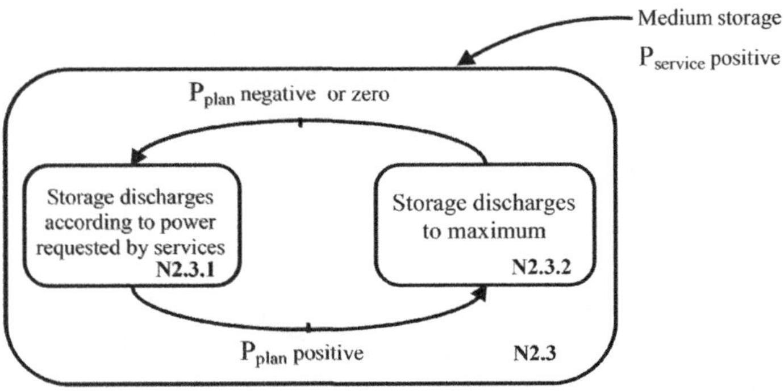

Figure 7.10. *Functional graph of mode N2.3*

N3.2 (Figure 7.11): the storage level is high and the power requested by services is positive or zero. To favor storage discharge, the supervisor gives a reference to discharge storage to maximum when the planned power is higher than or equal to zero. In another way, when the planned power is less than zero, the reference seeks to fulfill the power requested by services.

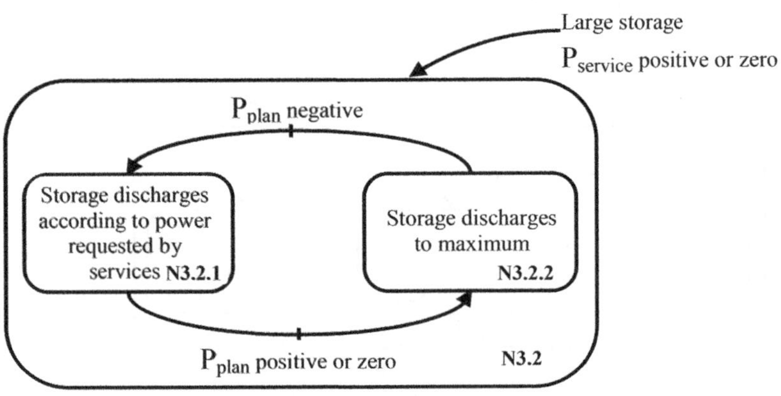

Figure 7.11. *Functional graph of mode N3.2*

The group formed by all of these operating modes is presented in the functional graph in Figure 7.12.

7.3.5. *Determination of membership functions*

The membership functions of the input values determine the transitions between the different operating modes and the value of the reference.

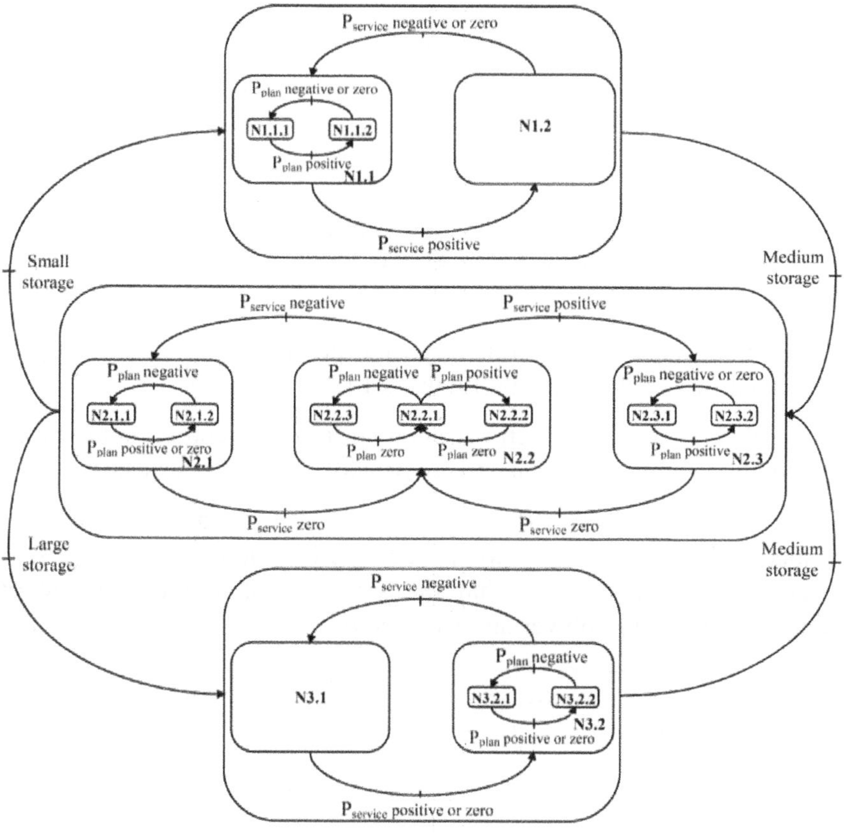

Figure 7.12. *Functional graph with all operating modes*

Membership functions related to storage levels (SOC) are composed of three levels (Figure 7.13), consistent with the three operating modes (N1, N2 and N3) of the previous graphic representation:

– sets “S” and “B”, for “small” and “big”, respectively, ensure the reserve energy necessary for the power plant to contribute to frequency control;

– set “M”, for “medium”, ensures the other objectives.

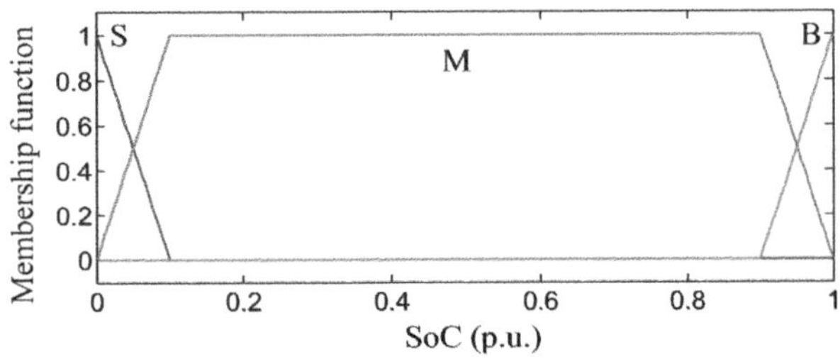

Figure 7.13. *Membership functions related to storage levels (SOC). For a color version of the figure, see www.iste.co.uk/robyns/powergrids.zip*

Membership functions related to the power requested by services are composed of three levels (Figure 7.14):

– set “Z” for zero, which is triangular in form, represents the mode in which storage is not requested to provide additional services;

– sets “NB” and “PB” for “negative big” and “positive big”, respectively, represent the power requested by additional services. A “negative” request means that the storage system must charge or reduce its discharge power and *vice versa*.

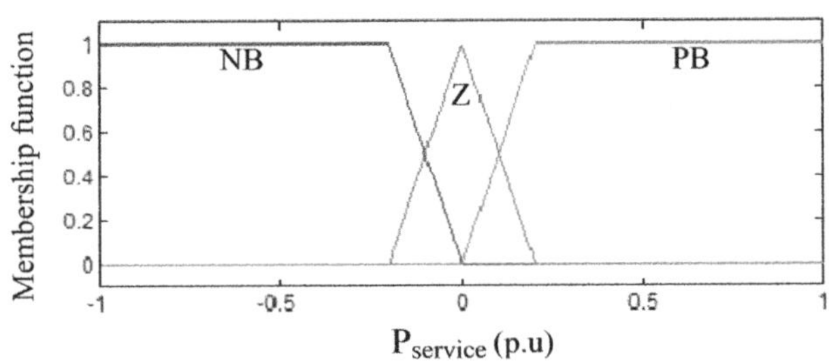

Figure 7.14. *Membership functions related to power requested by services. For a color version of the figure, see www.iste.co.uk/robyns/powergrids.zip*

Membership functions related to storage planned power are also composed of three levels (Figure 7.15):

– set "Z" for zero, which is triangular in form, represents the "standby" mode of the storage system. In this mode, the purchase/sale rate is not favorable for charge or discharge;

– sets "NB" and "PB" for "negative big" and "positive big", respectively.

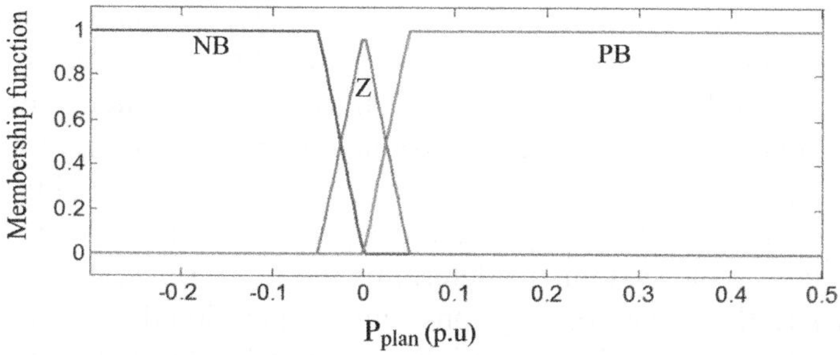

Figure 7.15. *Membership functions related to storage planned power. For a color version of the figure, see www.iste.co.uk/robyns/powergrids.zip*

For membership functions related to output (Figure 7.16), the reference power, five sets are considered to achieve a compromise between the precision of the power generated and the complexity of the supervisor. The sets are called "NB" (negative big), "NM" (negative medium), "Z", "PM" (positive medium) and "PB" (positive big).

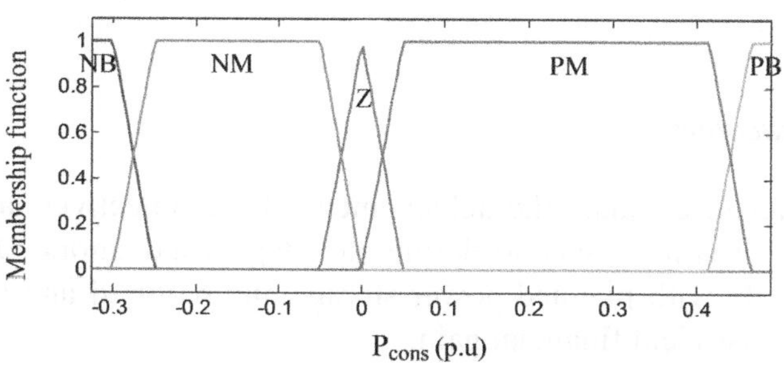

Figure 7.16. *Membership functions related to reference power. For a color version of the figure, see www.iste.co.uk/robyns/powergrids.zip*

The lack of symmetry in the membership functions related to reference power is due to the fact that the maximal charge and discharge powers are not identical.

7.3.6. *Determination of operational graphs*

The number of fuzzy rules associated with each output variable is determined by multiplying the numbers of fuzzy sets of each input variable with one another, or $3 \times 3 \times 3 = 27$. Traditionally, these rules are determined with the help of tables. The table associated with each output variable will, therefore, have four dimensions. The methodology proposed with the associated graphic representation has a dual advantage: it facilitates the writing of rules by avoiding the use of tables, and extracts only the most pertinent rules for the overall functioning of the system. To determine the fuzzy rules, it is necessary to translate the "functional graphs" into "operational graphs", which include the membership functions previously defined. Transitions between operational modes will be described by the membership functions of input values. Figure 7.17 shows the schema formed by all of the operational graphs.

7.3.7. *Extraction of fuzzy rules*

Fuzzy rules are directly extracted from the operational graph in Figure 7.17. They are shown in Table 7.2. Due to the methodology applied, only 13 fuzzy rules are considered instead of the 27 possible rules.

7.3.8. *Indicators*

In order to evaluate the achievement of the objectives targeted in Table 7.1, it is necessary to define the adapted indicators. Objectives having to do with planned power storage and required and additional services must yield financial gain.

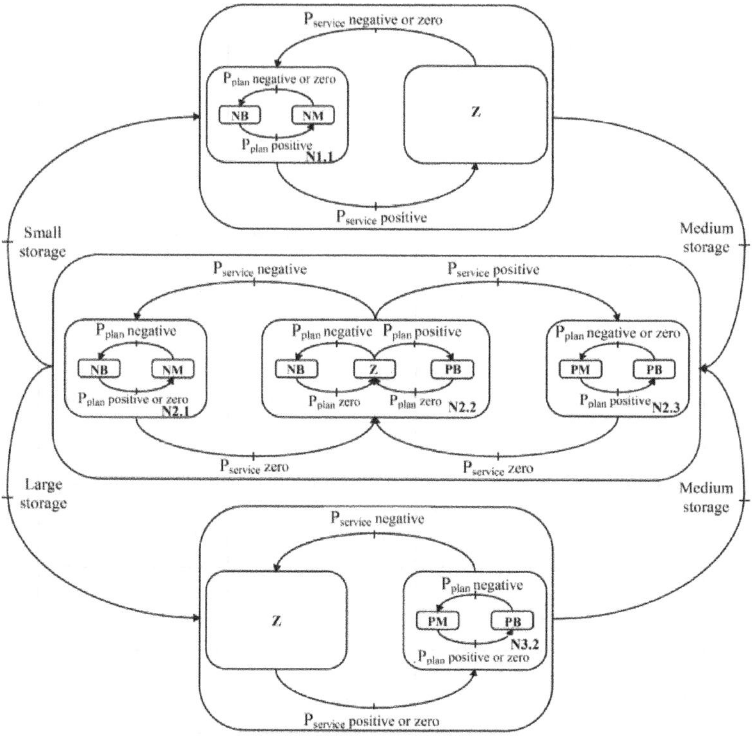

Figure 7.17. *Operational graph*

7.4. Economic value of services

The economic interest of a storage system can be obtained from three principal sources:

– purchase/sale action: buying electricity when the price is less expensive and reselling it at a higher price;

– supplying required ancillary services (frequency control);

– supplying additional services to the grid manager (congestion management) and to renewable producers (guarantee production).

The pricing details for these services will be presented in this section.

SoC	$P_{service}$	P_{plan}	$P_{setpoint}$
P	PB	PB, NB or Z	Z
P	NB or Z	P	NM
P	NB or Z	NB or Z	NB
M	Z	NB	NB
M	Z	Z	Z
M	Z	PB	PB
M	NB	NB	NB
M	NB	PB or Z	NM
M	PB	PB	PB
M	PB	NB or Z	PM
G	PB or Z	NB	PM
G	PB or Z	PB or Z	PB
G	NB	PB, NB or Z	Z

Table 7.2. *Fuzzy rules of supervisor*

7.4.1. Purchase/sale action

Purchasing and sales are billed based on the electricity price curve. When the storage system charges, the manager buys electricity, and the sale of stored electricity corresponds to the discharge phase. In this study, a price curve will be chosen with a dip during the night and two consumption peaks, one in the morning and one in the evening.

The price of electricity considered in this study as an example is shown in Figure 7.18. It is €15/MWh during off-peak time, and €40/MWh during peak time, and €30/MWh for the rest of the time.

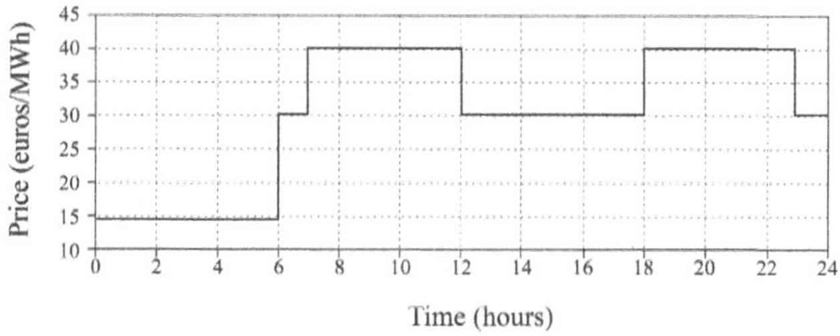

Figure 7.18. *Electricity price curve during 1 day*

7.4.2. *Frequency control billing*

According to the *Réseau de Transport de l'Electricité* (RTE) [RTE 13], remuneration for primary frequency control is €8.04/MW per half-hour step. Remuneration for secondary frequency control is composed of two terms; the first term corresponds to the maintenance of the reserve and is equal to €8.04/MW per half-hour step, while the second term corresponds to the use of the reserve and is equal to €9.30/MWh.

7.4.3. *Billing of additional services*

Additional services will be billed according to the contract signed between the parties. In this study, a rate of €25/MWh has been chosen; this is approximately equivalent to the gap between the high price (~€40/MWh) and the low price (~€15/MWh). This rate is the additional amount the storage supplier may receive (beyond the purchase/sale price) for each additional MWh it supplies or does not supply in comparison to the planned storage power. Because storage attempts to follow the demand for services, at least the same amount must be paid as the gain realized during a normal purchase/sale operation (40–15 = €25/MWh) between off-peak hours and peak hours.

7.5. Application

7.5.1. *Test grid*

The IEEE 14-node test grid [IEE 79] is composed of two different voltage levels (33 and 132 kV). The structure of the grid is shown in Figure 7.19.

The network is composed of 11 charges representing a total consumption of 259 MW and 73.5 MVAr. The generator at node 1 (132 kV) is separated into two in order to ensure the N-1 safety of the system. The power of each of the generators is 160 MW. The generator at node 2 has a power of 60 MW. Three synchronous compensators are connected at nodes 3, 6 and 8, respectively. Three wind farms are added at nodes 12, 13 and 14 at 33 kV with an installed power of 20, 50 and 70 MW, respectively [DO 12].

Compressed-air storage is installed at node 6. Its maximal power is 50 MW in discharge mode and 30 MW in charge mode, and its storage capacity is 500 MWh. Charge and discharge efficiency is estimated at 80%, giving an overall charge-discharge cycle efficiency of 64%, which is assumed to be constant for all operations in this study. The complete discharge time for the storage system is 10 h. The cavern storing air is supposed to have a volume of 160,000 m^3 with a maximum pressure of 30 bar and a minimum pressure of 20 bar. The maximum specific output of the air during charge is 50 kg/s and during discharge it is 120 kg/s.

Figure 7.20 shows the load curve for 1 day in per units (p.u) [RTE 12]. The load profiles are drawn from the typical variation during a day. The off-peak time of the night and two consumption peaks, one in the morning and one at night, are clearly visible.

The same wind profile is considered for the group of wind farms (Figure 7.21). This wind profile, adapted to the needs of the study, is highly variable, imposing significant variations in wind farm production. Line congestion occurs at 18 h (6:00 pm).

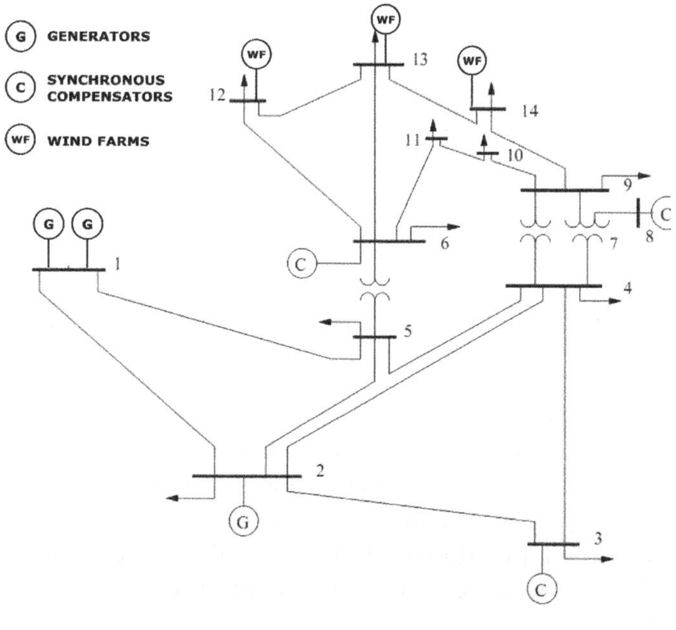

Figure 7.19. *IEEE 14-node network*

7.5.2. *Interest of the contribution of storage to ancillary services*

In this example is scheduled, we will compare the economic gain obtained when storage functions is scheduled according to planning alone with the gain obtained when it functions according to the real-time supervisor reference integrating planning.

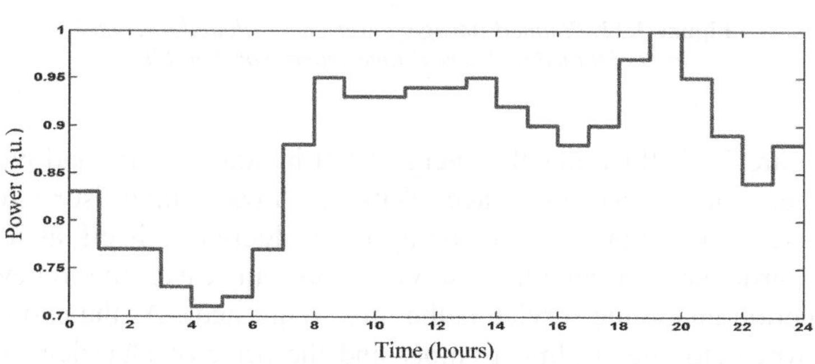

Figure 7.20. *Daily load consumption curve*

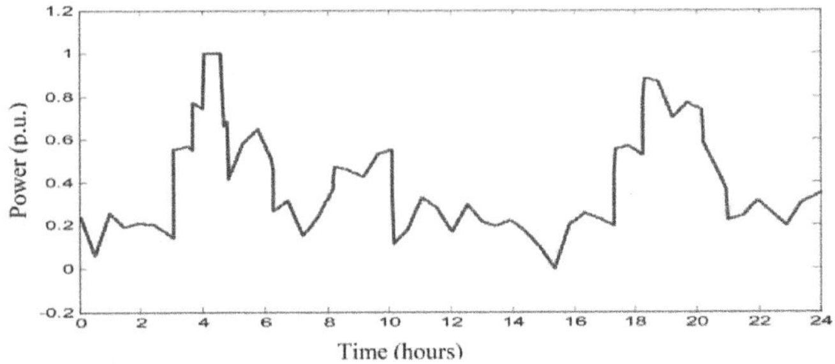

Figure 7.21. *Typical profile of normalized wind power over a 24-h period*

Figure 7.22 shows the storage power curve planned in the day ahead as a dotted line, and the curve obtained with a fuzzy supervisor as a solid line. We can see that without participation in ancillary services, the storage reserve margin for frequency control is unnecessary, and the storage system can discharge at its maximal power. The storage power curve with the fuzzy supervisor reference is more variable.

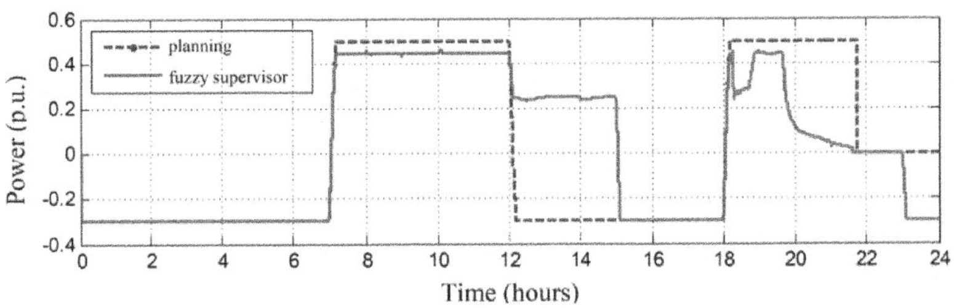

Figure 7.22. *Planned storage power curve (dotted line) and curve obtained with a real-time supervisor (solid line)*

Figure 7.23 illustrates the energy level in storage with and without the real-time supervision stage. With the fuzzy supervisor, storage discharges more than when planning is considered by itself, as it must also participate in ancillary services (in our case, the renewable production guarantee service is the most requested). At the end of the day, when storage is almost empty and the price of electricity is not too high, the fuzzy supervisor gives the storage reference to charge. For this reason, the state of charge of the storage system at the end of the day in this scenario with the fuzzy supervisor is higher compared to that which is obtained with planning alone. This difference is equal to 0.0536 p.u., which corresponds to a quantity of energy of 26.8 MWh. The energy statement for the storage system in both scenarios is given in Table 7.3.

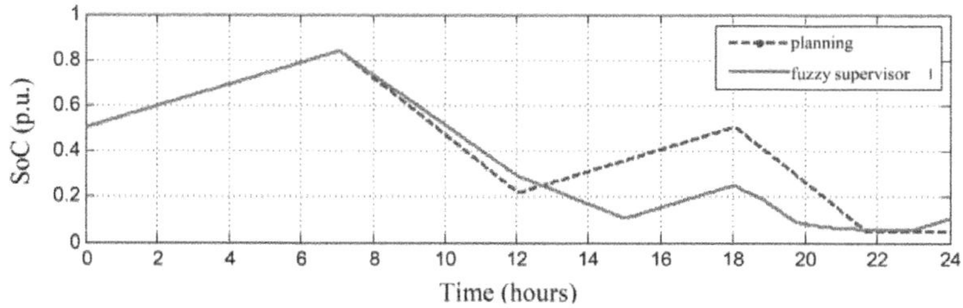

Figure 7.23. *Energy statement for storage obtained solely with planned power and with a real-time supervisor*

	Planning only	Planning and fuzzy real-time supervisor
Initial SoC	250 MWh	250 MWh
Charge	310.1 MWh	262 MWh
Discharge	535.1 MWh	460.2 MWh
SoC at the end of day	25 MWh	51.8 MWh

Table 7.3. *Energy statement for storage obtained solely with planned power and storage obtained with a real-time supervisor*

Figure 7.24 illustrates the evolution of financial gain with planned power alone and that obtained with the real-time supervisor. Table 7.4 shows the financial gains obtained per service over a span of 24 h.

The economic gain of the storage system over the day is larger if it functions according to the real-time supervisor setpoint. With the fuzzy supervisor, the storage manager can gain up to €11,700 at the end of the day, rather than €8,080 with planning alone, or a 44.8% additional gain.

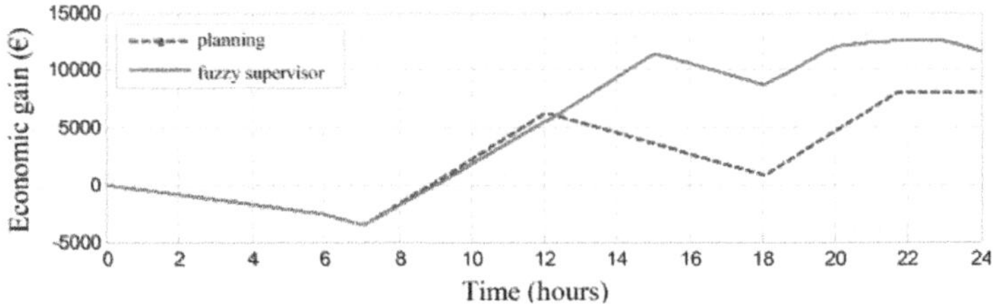

Figure 7.24. *Evolution of financial gain with planned power alone and that obtained with a real-time supervisor*

	Planning only	Planning and fuzzy real-time supervisor
Planned purchase/sale	€8,080	€6,840
Frequency control	–	€860
Additional services	–	€4,000
Total gain	€8,080	€11,700

Table 7.4. *Financial gain obtained with planned power only and that obtained with a real-time supervisor*

The economic gain of the purchase/sale operation is €6,840 with the fuzzy supervisor; lower than the €8,080 in the scenario in which the storage system functions according to planning alone. On the other hand, the energy level of the storage system at the end of the day is 26.8 MWh higher.

Though it is a required service of the storage system, the contribution of frequency control to the total economic gain is small; only €860 of the total €11,700.

The economic gain yielded by additional services plays a significant role in the overall whole: €4,000. This is the total of the remuneration the storage manager can receive for guaranteeing renewable production and congestion management.

Note that in this example, the absolute values of the gains inventoried are less important than their values compared between the two cases considered, which show the necessity of economic enhancing storage through multiple services. To determine the effective financial profitability of the storage system, it will of course be necessary to take into account the costs of amortization and maintenance over the functional life of the storage system (for example, 20 years).

7.5.3. *Interest of fuzzy supervisor compared to a Boolean supervisor*

In this section, the fuzzy supervisor will be compared to a Boolean supervisor. This Boolean supervisor follows the same rules as the fuzzy supervisor, but all the transitions between the states of the input and output variables are Boolean, as shown in Figure 7.25.

Storage power curves in the Boolean supervisor scenario and in the fuzzy supervisor scenario are shown in Figure 7.26. There is a visible difference between the two curves for the interval between 18 h (6:00 pm) and 24 h (12:00 midnight).

At 18 h (6:00 pm), there is congestion in one line. The grid manager requests the storage system to reduce its production in order to eliminate the congestion. With the fuzzy supervisor, the storage power curve is smoother. However, with the Boolean supervisor, this curve is highly variable. This can be explained by the fact that with the fuzzy supervisor there is a transition zone between the two states. Figure 7.27 shows a zoom of the storage power curve between 18 h (6:00 pm) and 19 h (7:00 pm).

At 20 h (8:00 pm), the state of charge of the storage system is low; the energy stored in the storage system is lower than 0.1 p.u. (Figure 7.28). With the fuzzy supervisor, the storage system reduces its output power in order to slow down discharge and conserve energy. This action can also be seen in Figure 7.28. The discharge of the storage system with the Boolean supervisor is, therefore, deeper than the discharge obtained with the fuzzy supervisor. The energy statement for the storage system in both scenarios is presented in Table 7.5.

From the perspective of economic indicators, there is no significant difference between the two variants of a real-time supervisor.

The latter case illustrates the impact of the membership function parameters initially determined empirically. These parameters can be optimized further in order to refine the maximization of gains.

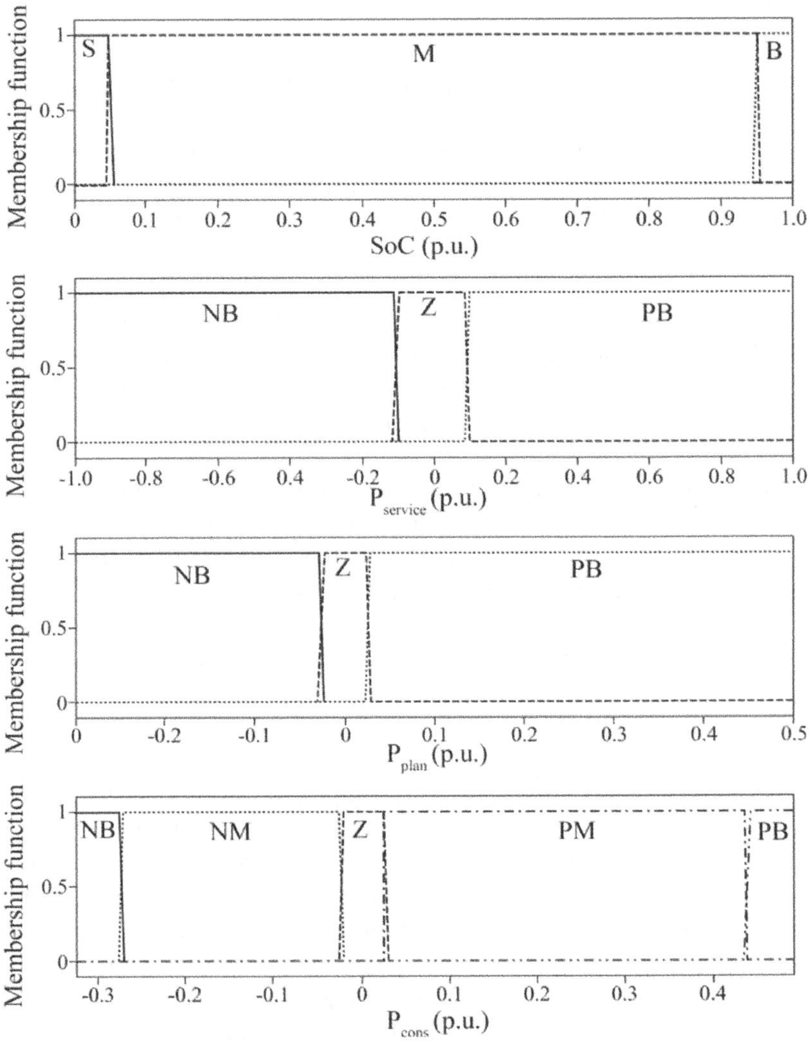

Figure 7.25. *Membership functions of a Boolean supervisor. For a color version of the figure, see www.iste.co.uk/robyns/powergrids.zip*

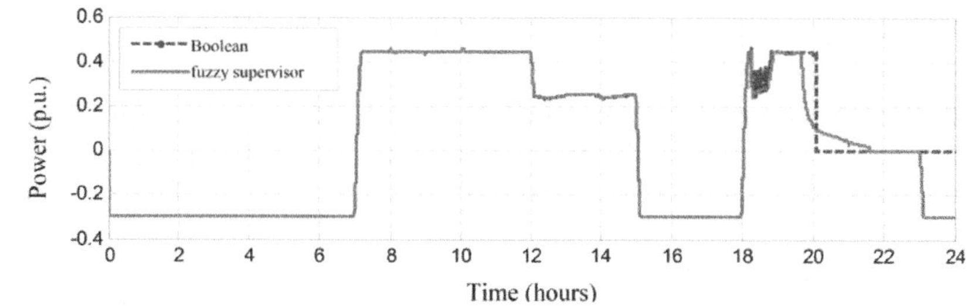

Figure 7.26. *Storage power curve in Boolean and fuzzy supervisor scenarios*

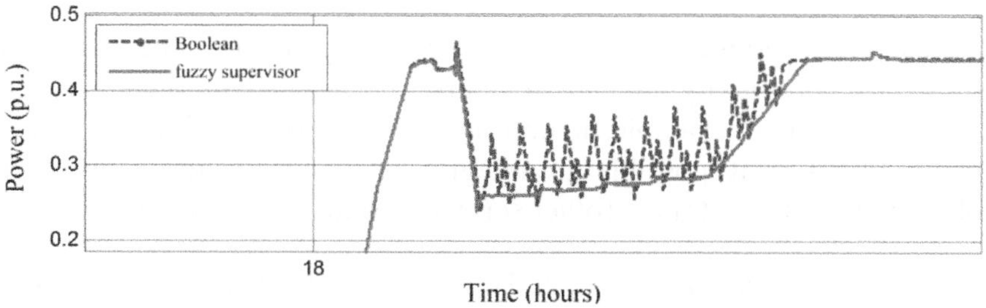

Figure 7.27. *Zoom of storage power at 18 h (6:00 pm)*

	Planning and Boolean real-time supervisor	Planning and fuzzy real-time supervisor
SoC initial	250 MWh	250 MWh
Charge	239.5 MWh	262 MWh
Discharge	464.5 MWh	460.2 MWh
SoC at the end of day	25 MWh	51.8 MWh

Table 7.5. *Energy statement for storage in Boolean and fuzzy supervisor scenarios*

7.6. Conclusion

In this chapter, a real-time supervision methodology making it possible to maximize the services provided and to contribute to the profitability of adiabatic CAES storage was introduced. A multi-objective real-time supervisor based on fuzzy logic has been developed to maximize the economic gain of the storage system, taking into account purchase/sale action and the services (required and additional) provided by the storage system, such as frequency control, congestion management and guaranteeing renewable production.

The supervisor developed has been tested on an IEEE 14-node test grid over a 24-h period. The simulation results have shown an economic gain for the storage system that is considerably greater if it participates in ancillary services and additional services, requiring real-time management.

The proposed methodology aids in the design of the supervisor and limits its complexity. It can of course be applied to other storage technologies (for example, hydraulic storage). The modularity of the method makes it possible to integrate it fairly easily with other objectives and constraints (other services, control of aging, etc.).

Finally, fuzzy logic, which is a determinist and non-stochastic method, generally presents a highly robust character with regard to uncertainties; particularly uncertainties involving the prediction of renewable energies. Supervisors constructed according to the proposed method have the capacity to adapt their responses to this type of hazard.

7.7. Acknowledgments

This study has benefited from the financial support of the French National Agency of Research (ANR) through the SACRE project, whose partners are EDF, Géostock, Laboratoires LMS, PROMES and L2EP.

7.8. Bibliography

[ACK 05] ACKERMANN T., *Wind Power in Power Systems*, John Wiley & Sons, New York, 2005.

[DO 12] DO-MINH T., Approche probabiliste pour l'évaluation de la fiabilité du système électrique intégrant des énergies renouvelables peu prévisibles, PhD Thesis, Université Lille 1, December, 2012.

[DO 14] DO-MINH T., MERDASSI A., ROBYNS B., "Gestion et valorisation d'un stockage à air comprimé adiabatique intégré dans un réseau électrique", *Symposium de Génie Electrique,* SGE '14, Cachan, 8–10 July 2014.

[IEE 79] IEEE, "IEEE Reliability Test System 1979", *IEEE Transactions on Power Apparatus and Systems*, vol. PAS-98, pp. 2047–2054, 1979.

[ROB 12] ROBYNS B., DAVIGNY A., BRUNO F. *et al.*, *Electric power generation from renewable sources,* ISTE, London and John Wiley & Sons, New York, 2012.

[RTE 12] RTE, Réseau de Transport d'Electricité, available at http://www.rte-france.com, 2012.

[RTE 13] RTE, Référentiel Technique de RTE, available at http://www.rte-france.com, 2013.

[VER 09] VERGNOL A., SPROOTEN J., ROBYNS B. *et al.*, "Gestion des congestions dans un réseau intégrant de l'énergie éolienne", *Revue 3EI,* no. 59, pp. 63–72, December 2009.

[VER 11] VERGNOL A., SPROOTEN J., ROBYNS B. *et al.*, "Line overload alleviation through corrective control in presence of wind energy", *Electric Power Systems Research*, Elsevier, vol. 81, pp. 1583–1591, July, 2011.

Index

Other titles from

iSTE

in

Electrical Engineering

2015

MEYNARD Thierry
Analysis and Design of Multicell DC/DC Converters using Vectorized Models

2014

BOILLOT Marc
Advanced Smart Grids for Distribution System Operators

2013

GALLOT-LAVALLÉE Olivier
Dielectric Materials and Electrostatic

GLAIZE Christian, GENIES Sylvie
Lithium Batteries and Other Electrochemical Storage

PERA Marie-Cécile, HISSEL Daniel, GUALOUS Hamid, TURPIN Christophe
Electrochemical Components

2012

BECKERS Benoit
Solar Energy at Urban Scale

FOULADGAR Javad
Electrothermics

GLAIZE Christian, GENIES Sylvie
Lead and Nickel Electrochemical Batteries

ROBOAM Xavier
Systemic Design Methodologies for Electrical Energy Systems

ROBOAM Xavier
Integrated Design by Optimization of Electrical Energy Systems

ROBYNS Benoît *et al.*
Electricity Production from Renewable Energies

SABONNADIÈRE Jean-Claude, HADJSAID Nourédine
Smart Grids

2011

GAO Fei, BLUNIER Benjamin, MIRAOUI Abdellatif
Proton Exchange Membrane Fuel Cells Modeling

SABONNADIÈRE Jean-Claude, HADJSAID Nourédine
Electrical Distribution Networks

LOUIS Jean-Paul
Control of Non-conventional Synchronous Motors

LOUIS Jean-Paul
Control of Synchronous Motors

MONMASSON Eric
Power Electronic Converters: PWM Strategies and Current Control

MULTON Bernard
Marine Renewable Energy Handbook

RAZIK Hubert
Handbook of Asynchronous Machine with Variable Speed

TRIGEASSOU Jean-Claude
Electrical Machines Diagnosis

ZAIM Mohammed El-Hadi, REZZOUG Abderrezak
Non-conventional Electrical Machines

2010

BERETTA Joseph
Automotive Electricity / Electric Drive Systems

BRUNET Yves
Energy Storage

DE FORNEL Bernard, LOUIS Jean-Paul
Electrical Actuators / Identification and Observation

JUFER Marcel
Electric Drive / Design Methodology

LE DOEUFF René, ZAÏM Mohammed El-Hadi
Rotating Electrical Machines

MARTINEZ-VEGA Juan
Dielectric Materials for Electrical Engineering

2009

HADJSAÏD Nouredine, SABONNADIÈRE Jean-Claude
Power Systems and Restructuring

HUSSON René
Control Methods for Electrical Machines

MOTTIER Patrick
LED for Lighting Applications

SABONNADIERE Jean-Claude
Low Emission Power Generation Technologies and Energy Management

SABONNADIERE Jean-Claude
Renewable Energy Technologies

2008

CRAPPE Michel
Electric Power Systems

MEUNIER Gérard
The Finite Element Method for Electromagnetic Modeling

2006

PRÉVÉ Christophe
Protection of Electrical Networks

Printed and bound by CPI Group (UK) Ltd, Croydon, CR0 4YY

13/09/2026

14961078-0004